MAIN LIBRARY
STO

**ACPL ITEM
DISCARDED**

338.47677 SH4M   7141774
SHELTON, CYNTHIA J., 1950-
THE MILLS OF MANAYUNK

ALLEN COUNTY PUBLIC LIBRARY

FORT WAYNE, INDIANA 46802

You may return this book to any agency, branch,
or bookmobile of the Allen County Public Library.

The Mills of Manayunk

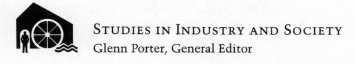 STUDIES IN INDUSTRY AND SOCIETY
Glenn Porter, General Editor

Published with the assistance of the Hagley Museum and Library

1. BURTON W. FOLSOM, JR.
*Urban Capitalists: Entrepreneurs and City Growth
in Pennsylvania's Lackawanna and Lehigh Regions, 1800–1920*

2. JOHN BODNAR
*Worker's World: Kinship, Community, and Protest
in an Industrial Society, 1900–1940*

3. PAUL F. PASKOFF
*Industrial Evolution: Organization, Structure, and Growth
of the Pennsylvania Iron Industry, 1750–1860*

4. DAVID A. HOUNSHELL
*From the American System to Mass Production, 1800–1932:
The Development of Manufacturing Technology in the United States*

5. CYNTHIA J. SHELTON
*The Mills of Manayunk: Industrialization and Social Conflict
in the Philadelphia Region, 1787–1837*

The Mills of Manayunk

Industrialization and Social Conflict in the
Philadelphia Region, 1787–1837

Cynthia J. Shelton

The Johns Hopkins University Press
BALTIMORE AND LONDON

This book has been brought to publication
with the generous assistance of the Hagley
Museum and Library.

Allen County Public Library
Ft. Wayne, Indiana

© 1986 The Johns Hopkins University Press
All rights reserved
Printed in the United States of America

The Johns Hopkins University Press
701 West 40th Street
Baltimore, Maryland 21211
The Johns Hopkins Press Ltd., London

The paper used in this publication meets the
minimum requirements of American National
Standard for Information Sciences—Permanence of
Paper for Printed Library Materials, ANSI
Z39.48-1984.

Library of Congress Cataloging-in-Publication Data

Shelton, Cynthia J., 1950–
  The mills of Manayunk.

  (Studies in industry and society)
  Revision of thesis (Ph. D.)—UCLA, 1982.
  Bibliography: p.
  Includes index.
  1. Textile industry—Pennsylvania—
Philadelphia—History. 2. Manayunk
(Philadelphia, PA.) I. Title.
II. Series.
HD9858.P5S5 1986   338.4'7677'00974811   86-7214
ISBN 0-8018-3208-X (alk. paper)

7141774

For Gary

Contents

	FOREWORD	IX
	ACKNOWLEDGMENTS	XI
	Introduction	1
I.	Labor and Capital in the Early Period of Manufacturing	7
II.	Textiles and the Urban Laborer, 1787–1820	26
III.	Mechanization and Mill Production in the Urban Seaport, 1820–1837	54
IV.	Economy and Society in Roxborough and Manayunk	76
V.	The Institutions of Order	100
VI.	Primitive Protest, Republican Reform, and Religious Revival	116
VII.	The Mill Strikes and the New Ideology of Class	134
VIII.	The Politics of Class Conflict in Early Industrial Society	156
	NOTES	175
	BIBLIOGRAPHIC ESSAY	211
	INDEX	217

Foreword

Cynthia Shelton's *The Mills of Manayunk* is a welcome and a most appropriate addition to this series. It adds substantially to our picture of the coming of industrialization to America. For many years historians focused on New England when studying the industrial revolution in this country. Recently we have had an outpouring of books dealing with the rise of the factory system and mass production in the Mid-Atlantic states, *The Mills of Manayunk* is an unusually important addition to that growing body of literature.

Shelton's analysis emphasizes the role of the immigrant British labor force that poured into the Philadelphia region in the late eighteenth and early nineteenth centuries. A plentiful supply of both skilled and unskilled labor shaped the evolution of early industry in the Pennsylvania metropolis. It was particularly important to textile manufacturers, who in Philadelphia turned to mechanization at a comparatively late point because of the availability of a large and relatively cheap work force. That same work force had come to the young United States imbued with the political economy of British trade unionism. When mechanization did reach the Philadelphia textile trade in the Manayunk mills, its introduction touched off unusually bitter social conflicts. The background and experiences of the immigrant workers made those clashes even more severe than they otherwise would have been. No other significant American textile community—either in New England or in the Mid-Atlantic states—appears to have undergone by the 1830s the degree of social stress that afflicted the Roxborough-Manayunk area in Philadelphia.

Using an impressive array of original sources, Dr. Shelton traces key elements of the history of early industrialization in the Philadelphia region in the half century following the writing of the Constitution. She begins with the experiments in manufacturing organized by John Nicholson along the Falls of the Schuylkill in the 1790s, a fascinating failure in which workers and owners took uncertain steps

Foreword

toward new economic and social relationships. The large-scale production of textiles undertaken by the Guardians of the Poor late in the eighteenth century eventually made Philadelphia a center for the manufacture of yarn and cloth, though generally not via the use of the latest water-powered machinery that marked New England's efforts. When large, mechanized textile factories did come to Philadelphia in the 1820s and 1830s, they arose along the Schuylkill River in the Manayunk area of Roxborough Township. Shelton offers a full portrait of the older Roxborough community of millers, farmers, and tradesmen in whose midst the new factories appeared. Local tax and church records permit her to sketch the complex religious, political, and social changes in Roxborough-Manayunk in the early decades of the nineteenth century. By the time of the depression in the late 1830s, a new social order was in place.

This book adds greatly to historians' understanding of the multifaceted face of early industrial America. It is a carefully crafted case study of one of the important but relatively little-studied centers of industry. Political, economic, and labor historians as well as scholars of social history will find *The Mills of Manayunk* a highly valuable piece in the puzzle of our early national history.

For all those reasons, and because Cynthia Shelton was a valued Fellow at the Hagley Museum and Library during the preparation of a part of this study, it gives me particular pleasure to have her work appear in this series. In varied discussions in Delaware, on Melville Street in Philadelphia, and in Los Angeles, I had the opportunity to offer my customary excellent editorial advice. Some of it was even taken, and the rest was listened to with good humor and grace. No reasonable series editor can ask for more.

GLENN PORTER, Director
Hagley Museum and Library

Acknowledgments

Historical scholarship is not a solitary endeavor, and I am pleased to acknowledge those who have contributed to this work. Two travel grants from the Philadelphia Center for Early American Studies, University of Pennsylvania, and a fellowship from the Regional Economic History Research Center, Hagley Museum and Library, allowed me to spend months with the sources in the Philadelphia region. The librarians and staffs of the Historical Society of Pennsylvania, Pennsylvania State Archives, Eleutherian Mills Historical Library, Archives of the City and County of Philadelphia, Historical Society of Montgomery County, and University Research Library, U.C.L.A., provided indispensable assistance in locating and identifying research materials. I am particularly grateful to Linda Stanley, manuscripts and archives curator at the HSP. She expertly and tirelessly pointed me to sources and unfailingly responded to my long-distance pleas for assistance. A special debt is also owed to my copy editor, Carlisle Rex-Waller, who skillfully tackled my manuscript with uncommon energy and patience.

It was my good fortune during research trips east to find room and board with historians Sharon Salinger, Michelle Maskiell, Billy G. Smith, David Dauer, and Steven Hahn. They shared not only their homes with me, but their knowledge of the past, of Philadelphia research sources and ethnic restaurants, and of the games of squash and tennis. I am further indebted to Sharon Salinger and Steven Hahn for making invaluable comments on key sections of this book.

An early draft of this work received a thoughtful reading from Joyce Appleby and Alexander Saxton of the department of history at U.C.L.A. I am grateful for their incisive critiques. As I made my way through successive revisions, I was fortunate to have access to a standing committee of tough critics in the Los Angeles Social History Study Group. Hal Barron, Jim Gregory, Susan Glenn, Roberto Calderon, Ronald Schultz, David Brundage, Frank Stricker, Peter Seixas, and Jim Prick-

Acknowledgments

ett provided important feedback on at least one of my chapters. I am particularly grateful to Jackie Greenberg, John Laslett, Steven Ross, and Devra Weber, the "oldies" of the study group, for their constant supply of hard criticism and warm friendship over the years.

I owe the greatest debt to three individuals. Mary Yeager supervised *The Mills of Manayunk* through its dissertation stage at U.C.L.A. She did so with a rigor that forced me continuously to rethink and refine my arguments. Her intellectual guidance has substantially strengthened this book. No one professed his or her faith in my work earlier or more strongly than Glenn Porter, editor of the Studies in Industry and Society series. From the beginning of this project, he has been a steady source of encouragement, thoughtful criticism, and solid advice. I am deeply grateful for his support.

The Mills of Manayunk have not been the center of my thoughts and interest over the past few years. Gary Nash has been. But I forgive him the distraction. For many hours he has put his mind to this work. His intellectual contributions cannot be itemized here. They can, however, be taken measure of in what may be found to be good about this book.

Introduction

This is a study of the emergence of the factory system and early industrial capitalist society in the urban Mid-Atlantic. The process began to unfold in the textile workhouses of the old commercial-manufacturing port of Philadelphia, where handlooms and wheels and their operators were gathered following the post-Revolution depression. The process culminated a half century later in the great water-powered, mechanized textile mills of Manayunk, in Roxborough Township, just six miles from the wharves of the port city. In 1824, a correspondent to the *Democratic Press* reported that "mill is huddled on mill" in the newly named borough of Manayunk on the Schuylkill Canal. Four years later, a town booster proclaimed Manayunk to be "the Manchester of America," drawing an ironic comparison to, not only the center of England's great mechanized cotton industry, but the home of a notoriously oppressive form of factory production and a militant trade union movement. It was then, in Manayunk, over the dozen years preceding the Panic of 1837, that the industrial wage-labor system was introduced to the urban Northeast. To a degree unmatched at the time in other manufacturing centers, social conflict and labor radicalism marked the early industrial history of Philadelphia.[1]

A major task of this study will be to show how deeply that history was bound to the experience of Britain's manufacturing laborers. Indeed, the key historical agent in my story is the immigrant textile worker. I am centrally concerned with recapturing the signficant role the immigrant male handloom weaver and mule spinner and the female powerloom weaver and throstle tender played in the social process that produced the industrial working and capitalist classes in antebellum America. That task draws me into a rich current of recent social histories on southern New England and the metropolitan Mid-Atlantic that have been concerned with the development of capitalist class relations in the early nineteenth-century Northeast. Focusing on the transition in the production and political relations of the urban tradesmen and rural industrial laboring classes, a number of historians in the last decade have shown that early industrialization was a complex but nevertheless discernable social process, whose unevenness was determined by historical and not accidental circumstances.[2]

Together these studies have illuminated the process of class formation in the antebellum period. They leave, however, a conceptual gap in our understanding of the development of industrial capitalist society, one that the history of Philadelphia's pre-1840 weavers and spinners can begin to fill. In focusing on the productive and political lives of textile workers, I hope to bring to view two aspects of the rise of industrial wage-labor production and the early industrial working class. First, the development of textile manufacturing in Philadelphia differed significantly from the mechanized textile districts of southern New England and produced sustained labor-capital conflict that appeared only sporadically in Lowell, Pawtucket, or other mill towns before 1840. Second, Philadelphia's immigrant weavers and spinners injected a radical analysis into the burgeoning working-class critique of the early industrial capitalist order.

A key to both developments was the transatlantic history of the immigrant textile labor force—its abundance, poverty, skills, and experience in production and protest in England as well as America. This influenced, in turn, the distinctive course that textile manufacturing followed in Philadelphia and the development of a working-class consciousness. Yarn and cloth in Philadelphia, unlike in New England, were manufactured in successive stages of handicraft, outwork, manufactory, and factory production between 1787 and 1850.[3] Philadelphia capitalists, moreover, turned to mechanized production of yarn and cloth later than their New England counterparts. British handloom weavers and mule spinners, who had been immigrating to the Delaware port since the 1780s, had turned Philadelphia into the center of fine yarn and cloth production by the time that the power-loom-based Waltham system was introduced to New England. Thereafter, immigrant women and children, disembarking in growing numbers in the 1820s, filled the demand for water-powered-machine attendants in the mechanized cotton and woolen factories of Manayunk. Philadelphia's position as an immigrant port in the post-revolutionary period was not shared by New England's coastal cities. Indeed, neither the Lowell mills nor those southern New England mill towns based on the family system of factory production drew upon English or Irish immigrant labor until after 1840. And no other mills created the exploitative factory system or the sustained strikes that Manayunk's did in the late 1820s and early 1830s.[4]

The experience and ideology of the immigrant textile laborer also

INTRODUCTION

are linked closely to the second aspect of Philadelphia's working-class history that this book explores: the development of labor-capital conflict and oppositional thought to early industrial capitalism. Many of the historians who have studied class formation in the early industrial Northeast share a central assumption about that process: it involved the transformation of the independent "republican artisan" of postrevolutionary America into a member of the dependent wage-laboring class. A peculiar political ideology and heritage, a unique set of material expectations and circumstances, and a particular cultural milieu had created the American mechanic in the preindustrial era.[5] By extension, these elements shaped his political response to industrialization—a process that ineluctably segmented and made a commodity of artisan labor and thus destroyed the economic and social status of the "producer" in one trade after another. We can infer from these studies that the ebb and flow of working-class consciousness and capitalist class relations and conflict in this period, whether in New York or Lynn, was controlled by the uneven proletarianization of the "American mechanic." Whether called "the mechanics' ideology" or "artisan republicanism," this peculiar world view was held by the generation of American-born craftsmen who came of age in the heady days of the postrevolutionary era.

In contrast, I will argue that the men and women of the textile trades of early nineteenth-century Philadelphia shared a different historical experience and an ideology distinct from that of the "American mechanic." They were not the sturdy castoffs from America's republican workshops but rather immigrant British textile operatives—"the eldest children of the Industrial Revolution."[6] They came of age in the textile mills of northern England and in the crowded and competitive handloom weaving trade of turn-of-the-century Philadelphia. The oppositional thought they brought to bear on their American circumstances derived much less from sentiments regarding equal rights imbibed by the American revolutionary generation than from English radical economic theory and trade unionism. In addition, weavers and spinners transmitted particular laboring-class oppositional strategies to Philadelphia, reflected in the machine breaking and strike movement that erupted in Manayunk and its factories.[7]

I came to this topic with assumptions about the course that Philadelphia's early industrialization would take and the role its weavers and spinners would play in that process. Marx's argument that the transformation of the labor process created fertile ground for class conflict underlies this study's close examination of the origins and process of mechanization in an old commercial handicraft center. E.

INTRODUCTION

P. Thompson's and John Foster's demonstrations of the breadth and political significance of that struggle in England's early nineteenth-century textile districts convinced me that it would be worth investigating the history of weavers, spinners, and textile manufacturers in early nineteenth-century Philadelphia and Manayunk. If factory operatives, "from the beginning . . . formed the nucleus of the Labour Movement" in England, as Frederick Engels argued in 1844, surely America's earliest factory operatives, who were British emigrants, must have contributed significantly to the history of labor's awakening in pre-1840 America.[8] Thirty years ago William Sullivan, in *The Industrial Worker in Pennsylvania, 1800–1840*, argued that class conflict and interest marked the development of factory textile production in the Mid-Atlantic. He revealed that Philadelphia's textile operatives were in the vanguard of the militant labor movement of the 1830s— a point overlooked by labor historians since. I hope to confirm his findings in this study of Manayunk's mill workers.

To explore the process of class development during the birth of the urban industrial economy brings one face to face with a historical moment of extraordinary popular activity and social and political conflict in Manayunk and Roxborough. In the decade preceding the Panic of 1837, immigrant factory operatives, industrial mill owners, tradesmen, farmers, small proprietors, and flour millers were swept into unprecedented alliances and campaigns in the form of mill strikes, political organizing, and religious revivals. Far from being coincidental paroxysms in separate arenas of popular action, such political and religious enthusiasm derived from a growing awareness of and contention over the social meaning of the factory. This was a time of particular tension, the dynamics of which were most succinctly captured by an Englishman grappling to define the transition he observed in his own society before the term "industrial revolution" was coined. "It is to this extraordinary revolution," John Wade wrote in 1833, that can be traced "the spread of an imperfect knowledge, that agitates and unsettles the old without having settled the new foundations."[9] Indeed, it was the "imperfect knowledge" of the social and economic impact of the factory system, of the relationship of immigrant factory worker and mill owner, and of the nature of authority in the mill community that underlay the struggle that was manifested in Roxborough's churches, political parties, and mills. A social order grounded in industrial capitalist class relations—"the new foundations"—was in the making.

This book begins a half century before the strikes, revivals, and political battles that occurred in Manayunk in the antebellum era. Industrialization was a long social process that originated in and transformed the relationships of the workplace. The story of that process thus starts with an examination of the unmechanized manufactories that John Nicholson built at the Falls of Schuylkill, a mile from Roxborough Township, in the 1790s. Nicholson represents the first generation of Philadelphia manufacturing capitalists who sought to organize labor and production in the promising market economy of the new nation. His workmen responded to and shaped the system of centralized wage-labor manufacturing. For Nicholson, his supervisors, and workmen, the adjustment to manufactory production was a fitful process in which the relations of capital and labor and the labor process itself were in flux. The nature of conflict between worker and owner was peculiar to this transitional stage of production in which the workman retained relative control over the production process. Nicholson's manufactories at the Falls of Schuylkill would give way, however, to a very different productive world in the mechanized factories of Manayunk.

Chapters 2 and 3 trace in broader fashion the experience of textile laborers as they moved from outwork, to shop, to manufactory, and finally to water-powered mill between 1787 and the early 1830s. This history of Philadelphia's developing textile industry and the regimen of the textile factory of the 1830s contrasts sharply with that of New England for reasons grounded in the experience of the immigrant weavers and spinners and their historical relationship with Philadelphia's textile capitalists.

Chapters 4 and 5 leave the workplace in order to analyze the changing nature of the social structure and social relations in the community of Roxborough as textile mills replaced flour mills as the base of the township's economy. The responses of Roxborough's tradesmen, laborers, farmers, and proprietors to the social and economic imperatives of factory society, manifested in the revivals and organization of local parties, were grounded in an earlier era, when merchant flour millers and farmers composed a paternalistic elite in Roxborough.

The dual concerns of the first five chapters, the transformations in the world of production and the concomitant change in the social and economic relations of the industrializing community, set the stage for the unprecedented political and cultural conflict that occurred in Roxborough and Manayunk in the late 1820s and 1830s. This is the focus of the remaining chapters. Chapter 6 examines the social basis of the religious revivals and the rise of the Workingmen's, local Jacksonian, and National Republican parties between 1828 and 1833. Chapter 7 investigates the significance of the strikes of 1833–34 and the birth of the trade union movement among factory operatives in

INTRODUCTION

Manayunk and the other textile districts of Philadelphia. The transatlantic history of Manayunk's mill operatives underlay a powerful attack articulated in the local newspaper against the conditions and relations of industrial capitalist production. The final chapter traces the political awakening on the eve of the Panic of 1837 of those outside the mills to the new relations of class, born behind the mill gate, that had come to divide their community.

I

Labor and Capital in the Early Period of Manufacturing

In 1794 John Nicholson, a wealthy Philadelphia entrepreneur and speculator, established a manufacturing village at the Falls of Schuylkill, less than a mile from the mouth of Wissahickon Creek and Roxborough Township. The enterprise failed within four years, yet the extant records of the short-lived venture open the door to the little known world of the non-mechanized factory in the take-off years of domestic manufacture. It is appropriate to begin the story of the coming of early industrial capitalism inside Nicholson's workshop. The post-1820 mechanized factories did not emerge full-blown on the banks of the Schuylkill but developed, in part, from the experiences of Philadelphia workers and capitalists in the nonmechanized manufactories. By observing the social relations within the manufactory of the 1790s, we can learn how employers and employees strove to fashion the system of centralized production and to adjust to each other as the first generation of wage earners and capitalist manufacturers.

Philadelphia entrepreneurs undertook their first ventures into domestic manufacturing in the 1790s imbued with capitalist expectations and republican visions. New opportunities and markets opened as the postwar commercial crisis ended. At the same time, pervasive urban unemployment and poverty forced republican idealists to reassess America's supposed escape from the extremes of economic inequality that plagued European society. Many of the Philadelphia-based nation-builders became convinced that domestic manufacturing would be the only way to maintain a self-sufficient, fully employed, and thus virtuous citizenry in the expanding postrevolutionary society. In 1787 Tench Coxe led fellow Philadelphia manufacturing advocates in the establishment of the Pennsylvania Society for the Encouragement of Manufacture and the Useful Arts (PSEMUA). Joining Coxe as a founding member of its committee to promote manufactures was John Nicholson.[1]

A young Philadelphian of thirty-four when he began plans to manufacture glass, yarn, buttons, and hosiery, Nicholson already had made and lost thousands of dollars in the volatile postrevolutionary land market. A speculator par excellence in the 1790s, he controlled on paper over four million acres of Pennsylvania land when he died in debtors prison in 1800. Nicholson and his partner Robert Morris had engaged in too many unsound land schemes and spread their credit too thinly. Nicholson's Schuylkill manufacturing complex was one of many investment schemes he engineered to salvage failed land deals. As an early member of the corps of Federalist manufacturing promoters who revolved around Hamilton and Coxe of the Treasury Department, Nicholson had a visionary enthusiasm for home manufacture. With the assurance that he would gain both the "merited esteem of [his] country" and "advantage in a pecuniary way," Nicholson began planning an ambitious manufacturing experiment in the spring of 1793.[2]

Support and advice for Nicholson's venture came from European mechanic-manufacturers with whom he made contact and who shared his belief that prospects for profitable manufacturing in America were at hand. These men had an ambiguous role in the early stage of factory production, for they were partly skilled workmen, partly supervisors, and partly owners. Their skills and knowledge of production were indispensable to capitalists like Nicholson, who had no experience in manufacturing and in managing labor.

Consulting with a handful of English, Scottish, and American machine builders and mechanics, Nicholson decided not only to launch a spinning and hosiery manufactory but to establish a diversified and self-contained manufacturing complex with his advisors as partners and managers. William Pollard, who had acquired the first patent in Philadelphia for the Arkwright spinning frame, agreed to build and manage the cotton mill. Nicholson persuaded John Campbell to abandon Alexander Hamilton's struggling project at Paterson, New Jersey, and convinced John Lithgow and William England to leave Scotland and build stocking frames and supervise hosiery production for him. Two English glass manufacturers advised him on the glass bottle business. Nicholson hired Charles Taylor, builder of the Albion Mills of England, to supervise the making of steam engines.[3]

These men, drawing on the models of English factory communities, outlined a plan to turn Nicholson's three hundred acres at the Falls of Schuylkill into a "handsome thriving little town." Pollard detailed for Nicholson the size and arrangement of tradesmen housing to be placed next to the mills at the Falls. Taylor suggested buying up existing housing "as it looks over immediately to the factory whereby we can have our Victuals and return without the loss of much time."[4]

To establish and operate the Falls farm, which would complete the "valuable little village," Nicholson brought a husbandman, George Stumme, from Germany.[5] When the French traveler Rochefoucauld-Liancourt visited the "extremely well chosen" site in the spring of 1795, he was impressed with the unfinished complex of warehouses, foundry, glass and stocking manufactories, workmen's dwellings, and company store. The various buildings spanned both sides of the Schuylkill, and Rochefoucauld-Liancourt noted that a particular building was assigned to each branch of labor. He was particularly impressed that the largest building was designed "for the habitation of the workmen, of whom Mr. Nicholson will be obliged to keep at least a hundred." "Everything promises success to the undertaking," the French visitor wrote.[6]

One element, however, clouded the optimistic forecast for Nicholson's undertakings. "Before any person begins manufacturing," the Scottish loom builder John Lithgow warned Nicholson in the fall of 1794, "he should consider how he is to be furnished with these three articles—Machinery, Materials and Workmen."[7] To secure workmen for a manufacturing labor force was difficult and expensive in late eighteenth-century Philadelphia. Through the 1790s, the complaint that "labor is so high" or "manual labor is so expensive" typified what manufacturing entrepreneurs viewed as their primary obstacle.[8]

Manufacturers like Nicholson confronted a tight market for unskilled workmen such as carters, boatmen, button makers, and quarrymen during the season when hinterland farmers demanded harvest workers. Rochefoucauld-Liancourt summed up Nicholson's predicament: "There are few good workmen who are with difficulty obtained, and whose wages are exorbitant."[9] In a country where there was not even a reserve of unskilled labor, skilled tradesmen, including glassblowers, framesmiths, and machine builders, were particularly difficult to find and expensive to hire. The possibility of substituting machines for laborers was severely limited in this period. Machine technology and mechanization had not been developed to replace human skills in glassmaking, stocking weaving, and fine yarn spinning. Labor-saving innovations, which converted unskilled labor into cheap skilled labor, had been introduced in coarse yarn spinning, which was to be the centerpiece of Nicholson's venture. But even here Nicholson faced stiff competition for the skilled mechanics to build the Arkwright cotton-spinning frame.[10]

Early American textile capitalists like Nicholson encountered a particular hardship in Britain's protection of its monopoly of skilled textile operatives and machine builders. Parliament had passed stiff laws in an effort to prevent both the emigration of mechanics and export of machine models. To compete successfully, manufacturers developed stealthy tactics to evade the laws and vie for British frame builders and engine designers. From the beginning, Nicholson and his textile experts discussed, plotted, and played the game of commercial subterfuge as readily as other American manufacturers.

One of the first concerns of Nicholson and his partners was how "to obtain artists" from England and Ireland. Nicholson was well aware that they would have to take care "to prevent the interposition of government." Pollard, who had carefully considered the problem of acquiring skilled labor, was optimistic that they "should not be at a loss for workmen in [their] present Plan." The stiff labor market in England, Scotland, and Ireland, he felt, would induce "unfortunate" competitors to immigrate to America.[11]

John Lithgow did not share Pollard's optimism about the market for skilled labor either in Philadelphia or Great Britain. He argued that Philadelphia "workmen are next to impossible to be met with . . . and as to getting them from Great Britain in sufficient numbers the idea is absurd, and would be unprofitable." Lithgow urged Nicholson to tap the rural labor market. "In any Seaport Town," he explained, "it would be difficult to get textile workers at a reasonable wage." Laborers received a dollar per day in Philadelphia. "To manufacture stockings or weaving of any kind in Philadelphia," Lithgow wryly commented, "is like building ships at Pittsburgh." If textiles were to prove profitable, the manufacturer would have to locate in an inland town "where men will be satisfied with five shillings per day."[12]

Lithgow's pessimism about the labor market was warranted, as Pollard and John Campbell soon discovered when they began construction of textile machinery. Within a month, Pollard was complaining of the difficulty "in getting Workmen and at advanced prices to what I expected." He found that he could not employ workers to labor for six shillings (seventy-eight cents) per day, as he had anticipated. He also discovered that he had to compete with farmers for workmen in the warm months. Because tradesmen were "scarce to be got," construction of the spinning mill was delayed for weeks. Pollard also complained that his work was "all new" to American workers. "Much disappointed by [American] tradesmen," Pollard personally contacted skilled British spinners and machinists and urged Nicholson to raise wages above the rate in New York and Paterson and to pay promptly.[13] Campbell increased the wages of his frame builders to over two dollars per day in order to compete for workers and illegally attempted to bring framesmiths from New York.[14]

Facing stiff competition for textile machine builders and operatives, Nicholson also found that glassblowers in America were in short supply and that they commanded high wages, earning a dollar per day by the end of 1795.[15] In order to procure a labor force for his glass house, he advanced wages and passage from Amsterdam to a number of German glassblowers and members of their families. Six of one such family, from three generations, were employed in Nicholson's newly completed glass manufactory. Nicholson personally negotiated to bring a company of five German-born, Baltimore glassmakers to the Falls of Schuylkill. The Baltimore team assured Nicholson, "If you should be in want of good hands, we . . . think ourselves capable, to do the Business to perfection." They demanded wages of thirty dollars per month or half of that and a "free house included," reminding Nicholson that they had "a very good prospect that several of us will get Employment somewhere else."[16] Nicholson subsequently contracted with three of the five glassmakers on their terms. He arranged housing for two and paid the passage from Germany of the wife of one of the men. Within a few months of their employment, however, the Baltimore glassblowers threatened to leave Nicholson, apparently "bribed by some people of Jersey."[17]

The labor market for unskilled workers differed significantly from that for skilled. Nicholson was not compelled to look to Germany or England for button workers, boatmen, or quarrymen, who commanded one-half to three-quarters of the wage rate of glassblowers, framesmiths, or machinemakers. Yet even unskilled labor became scarce during three or four months of the year when he had to compete for workmen with local farmers.[18]

Nicholson experienced first hand the difficulties of securing a cheap labor force in the vicinity of the port city even before he began manufacturing. In the interest of increasing the value of his landholdings at the Falls of Schuylkill and in the interior of the state, Nicholson had invested in the construction of the Delaware and Schuylkill Canal and the Lancaster-Philadelphia Turnpike. The Delaware and Schuylkill Navigation Company was incorporated to connect the two rivers above the Falls and to open a water route to Reading. As a member of the board of managers, Nicholson demonstrated his business acumen, leading an effort to control wage rates of the construction workers. He proposed that the canal company form a combination agreement with local employers of day laborers to keep wages down. A committee conferred with other canal companies and the turnpike company and was given "powers to limit prices beyond which the respective companies will not give the Labourers they may employ other than those expected from New England in order that no competition of prices may occur."[19]

After the first year of construction on the canal, the supervisors struggled to hold down the cost of labor. The account books of the Delaware and Schuylkill Company reveal that very low daily wages were paid out through May 1794. Regular workers received three shillings, or forty cents, per day. But in the summer of 1794 competition from local farmers for harvest workers almost doubled wages to between sixty-six and seventy-three cents per day. Facing a shortage of operating capital as winter set in, the company moved to lower wages and stabilize the labor supply. The supervisors laid off all the day laborers and rehired them with a new contract. From that date forward, no worker was employed except by contract, and the wages were reduced to five shillings per day. Only during the harvest season were low-paid canal workers able to procure a raise in their daily pay, back up to five shillings, six pence.[20]

Nicholson also learned that he had to compete with local farmers for the labor of his quarrymen, who were digging stone for his canal and turnpike company. In January 1793 William Pollard, who supervised the quarry for the turnpike project, complained to Nicholson that the demand for farm laborers, who could command thirty-two cents per day plus board, was driving wage rates up. Pollard was forced to offer quarrymen at least forty-eight cents because he could not get accommodations for them. He then contracted with farmers in the area to bring in companies of workers for the quarry and supplemented this work force with "struggling hands from town."[21] Like the skilled workers who later came into Nicholson's employ, the quarrymen continually pursued the highest pay in the area. Within the year Pollard complained that the most experienced of his crew had left when Nicholson's turnpike company failed to meet competitive wage rates.

Faced with unexpected difficulties in recruiting a work force, Nicholson also confronted problems with the discipline of those he hired. He and his manufactory supervisors attempted to organize and to manage production effectively and profitably. Yet their attempts were undermined, they judged, by the customary work habits and attitudes of those who labored for them.

Nicholson's diverse manufacturing complex made both the tasks of balancing accounts and supervising labor time consuming and difficult. No clear distinction emerged between owner and manager in this early system of factory management. Nicholson was at the top of a hierarchy that oversaw production and marketing of the various branches of the complex and was involved almost daily in operations. He set up each manufactory as a department, headed by a superintendent who managed the workers and was required to attend weekly meetings. A manager of the store at the Falls handled the distribution of the manufactured goods and the sale of materials and staples to the department heads and workers. Nicholson required an agent at the Falls, Henry Elouis, to oversee the transportation of manufactured goods and materials between the various manufactories, the store, and Philadelphia and to send daily reports.

Nicholson made clear that careful accounting procedures were necessary "so as to give a view of the Profit of the work." He required the store manager to keep an order book, day book, journal, and ledger to record all orders for products and sales. Each department head received explicit instructions from Nicholson to deliver their manufactured goods to the store and to "keep a book like a Bank book" so the store manager could record their deliveries. Nicholson believed that meticulous bookkeeping would facilitate the exchange of "information and advice" and help "produce the greatest profit."[22]

To facilitate control of the labor force, Nicholson gave each superintendent authority over his workers. "I trust you will do as far as you can," he confided to one of his managers, "to see that I do not pay more than I ought—and that I am faithfully served by those I employ." To enable the supervisors to keep "an Eye to [the workers] dayley," Nicholson provided lodging for them at the Falls. The quarry superintendent enthusiastically endorsed this arrangement: "I could every moment [see] how the business there would be done."[23] Nicholson facilitated the face-to-face control of his supervisors by authorizing them to dispense housing to their charges and by making himself visible to his workmen, traveling to the Falls complex regularly from Philadelphia to see and be seen by his wage earners.

Despite this careful plan of management and supervision, the manufactories did not produce profits for Nicholson and his supervisor-partners. For this they blamed the behavior of their laborers. Not only did the workmen resist attempts to structure and standardize the work day, but they were not as productive as Nicholson and his supervisors had hoped. Rochefoucauld-Liancourt praised Nicholson's choice of "very able men" to conduct the manufactories but was pessimistic about the labor force: "A whole year may elapse, before the workmen fall into a proper train of business." This circumstance, he warned, did "not afford the most flattering prospects of success."[24] To prove Rochefoucauld-Liancourt's assertion, the Falls supervisors could have compiled a list of their complaints about their employees: the refusal to accept job assignments, unwillingness to conform to a timetable or structured work regimen, low productivity, and disregard for property.

Evidence from the supervisors' correspondence indicates that individual workmen refused certain tasks. Henry Elouis, the quarry superintendent, for example, angrily requested permission to discharge an unskilled flatman and general laborer who did not "like to be Employed at all kinds of work." At the glass manufactory, William Eichbaum could find no one to be a teaser, the arduous job of stoking the furnace fires.[25]

The work habits of the employees constantly frustrated the supervisors. Both the unskilled and skilled workers periodically took off from work, sometimes for days at a time. The transportation of materials and manufactured goods was recurrently interrupted by the "intolerable behavior" of teamsters and boatmen who abandoned their loads. Boatmen Patrick Doad and Samuel Right caused a "horrid delay" when they chose to go into Philadelphia for two days "to see some of their relations come from Ireland with the Morning Star." In the winter of 1796, Henry Elouis reported to Nicholson that some of his quarrymen and laborers were "so idle that they cannot be discharged too soon." The young unskilled assistants in the glass manufactory would not be disciplined either. Eichbaum complained of the boy attendants who preferred to "employ their time in fishing instead of working."[26]

Such irregular work schedules often contributed to expensive delays or loss of goods. William Pollard blamed the postponement of his plan to get cotton yarn on the market on a whitesmith who simply abandoned the spinning machine he was building. The production pattern of the glassmakers caused the most costly damages. Glass manufacturing did not conform to a prescribed working day. The furnaces, prepared during the day, demanded constant tending, and a batch of glass had to be worked as soon as it was ready, usually during the cool of the night hours. During one month Eichbaum charged some of his glassmakers with lost time for leaving the manufactory with "the best glass in the fire and the greatest demands." On another occasion, the glassmakers so frustrated Nicholson for "burning the coals and making no glass" that he threatened to jail them.[27]

From the managers' point of view, manufactory workers frequently did not comply with a standard work schedule because of drinking, a customary practice of eighteenth-century work life. "Country rum" and Lisbon wine were standard supplies at the Falls store, and Nicholson occasionally met direct requests for spirits, judging it an encouragement to productivity. But supervisors regularly sent complaints to Nicholson that intoxicated and disorderly glassmakers, quarrymen, boatmen, carpenters, and other employees had brought production to a standstill. A winter drinking binge by the glass workers, who traditionally consumed large amounts of water or liquor because of the intense heat of the fires, halted glass production for days. William Eichbaum reported to Nicholson that during one afternoon most of his workers became intoxicated, preventing the nightly pot-setting. "Skirmishes proceeded from Drunkenness" and drinking continued through the next day. Eichbaum informed Nicholson that he would not be able to see him in Philadelphia until "Bacchus' Reigne is over."[28]

Beyond being dismayed at the idleness and intemperance of the workmen, Nicholson and the supervisors basically mistrusted them. As far as Henry Elouis was concerned, the Falls "was surrounded by a good many rogues" who would just as soon "rob and abuse" Nicholson. The workmen had prompted this comment when they cut down trees and split boards on Nicholson's property. Even the store manager was skeptical of the workmen. He placed locks on the doors of the store, assuming that they would steal any goods that were left unprotected.[29]

Labor thus appeared to present a twofold problem for the early capitalist manufacturer: it was expensive and unreliable. Unskilled button workers and quarrymen were relatively inexpensive at approximately fifty-four to seventy-eight cents per day, but skilled glassmakers, foundrymen, cotton spinners, whitesmiths, and stockingmakers commanded no less than a dollar for a day's work. Moreover, these costly workmen, according to their supervisors, were not productive wage laborers. Henry Elouis's complaint that "a steady and sober man is much wanted" expressed the frustration of the supervisors in every manufactory.[30]

The profit-minded investor and supervisors portrayed an overpriced and unmanageable labor force prone to idleness, drunkenness, and disregard for property. The analysis of the character of early factory laborers and of the relations of labor and capital, however, must be taken further. When we begin to examine the interest and position of the wage laborers, the factors that defined the social relations of production and inhibited the success of capitalist manufacturing in this period appear much more complex. Nicholson's labor problems were not just a case of what Herbert Gutman has called the first-generation factory workers' "irregular and undisciplined work patterns [that] frustrated cost-conscious manufacturers."[31] Nicholson's workers were neither unable nor unwilling, as E. P. Thompson and Gutman might suggest, to engage in a structured wage-labor system of production.

To explain the confrontation between labor and capital in this period, it is necessary to understand the labor process and relations of production within the manufactory and how they might have changed.

The system of production in Nicholson's various manufactories resembled what Marx has defined as manufacture in its simplest form: the simultaneous employment by one capitalist of a number of wage-earning craftsmen who do the same kind of work. The gathering of workers under one roof gave the employer a control over production and the length of the working day that was impossible without centralization. Managers became necessary to enforce the prescribed hours of labor. This differed from handicraft production, which was generally individual in character. Whether he was a propertyless journeyman or independent master, the handicraft producer generally worked at his own pace and in his own fashion.[32]

During the initial transition to centralized production, the labor process of the bottlemakers, framemakers, and stocking weavers did not change. This is a crucial distinction to make between the early manufactory—what Sean Wilentz calls the "machineless factory"—and the mechanized factory. Nicholson achieved a division of labor in buttonmaking, but the glassmakers, framesmiths, and stocking weavers maintained the traditional processes of their trade within the system of capitalist manufacture. Nor were Nicholson's employees unfamiliar with the exchange of labor power for some type of wage payment. Journeymen in traditional handicrafts had frequently received direct wages from their masters. What had been lost in the transition form handicraft production to manufacturing, however, was the opportunity available to journeymen, stocking weavers, and framesmiths to set themselves up as independent masters.[33] The interest of those who labored in Nicholson's manufactory sprang from their dual position as skilled workmen who retained control over the labor process and as wage earners who relied on their wage payments for subsistence.

The conflict between Nicholson and his laborers arose from the inconsistencies of the early factory wage-labor system. While Nicholson, the capitalist manufacturer, bought, organized, and supervised the labor power of his hired workmen, he failed to pay money wages in a consistent or systematic way. The irregular and unproductive work habits described by the supervisors occurred within the context of unreliable wage payments. Far from proving that the workers clung to accustomed patterns of work time and stubbornly resisted a new system of labor, the drinking binges, the impulsive break from work to visit relatives, and the disregard for the product of their labor can be seen as early forms of labor protest. Customary practices of precapitalist work life—the daily breaks for grog in the shop and the irregular pattern of seasonal and custom production—certainly provided the basis for this type of activity. These actions, however, can also be interpreted as part of a broad range of individual and collective opposition to an uncertain system of wage payments. However high wages were in Nicholson's view, they were the vital means of subsistence for his employees and hence needed to be paid regularly. Nicholson's workmen also voiced their protests and took positive steps to guarantee that the wages they earned were paid promptly.[34]

Within months of starting production at the Falls complex, Nicholson was failing to meet the routine demands of the various branches for both weekly wage payments and materials. Despite the fact that "the want of [his manufactories'] commodities . . . ensure[d] them a certain market," Nicholson did not have enough operating or liquid capital. Engaged in so many manufactories at the same time that he faced large debts for failed land investments, he had spread his capital too thin. While he complained in early 1795 that the stocking weavers "are not affording me much profit," he knew well the circular nature of his dilemma: if he did not supply cash to his manufactories, they would not produce a profitable return.[35] As owner-manager, Nicholson realized that a sound wage system was essential to the profitable operation of his enterprise. Over the next two years, the workmen, supervisors, and Nicholson struggled over the issue of regular production and regular wages.

The workmen adopted a number of strategies to obtain steady wage payments. Some workers simply left in search of a more reliable employer. Groups of the general workmen at the Falls village—the carpenters, quarrymen, and boatmen—quit in waves when intermittent wage payments became a chronic problem. Nicholson had to replace his carpenters and joiners in early 1795. When the carpenters warned the disgruntled quarrymen that Nicholson would not pay any wages, they left their jobs.[36] By the summer, most of the original quarrymen were gone. Nicholson found a new company of quarrymen who were willing to accept a piece rate to replace those who had quit in July. But within a few months they, along with the boatmen and remaining carpenters, were "highly discontented" over not receiving their wages. The local butcher refused to give meat to the quarrymen, who could not pay their bills. And the carpenters complained that they could not get materials with "nobody willing to trust them." By the spring of 1796, the supervisor of the general laborers reported to Nicholson that he did not expect any of them to stay for lack of pay.

The more highly skilled employees stuck through months of irregular wage payments because they believed they could compel Nicholson to meet their demands. The whitesmiths weathered weeks without pay by drawing on communal support and the sympathy and support of their supervisor, William Pollard. By early 1795, Pollard's efforts to complete the spinning machinery for the Falls mill was slowed by lack of cash to pay wages. In early March, Pollard's workmen had not received wages for six weeks and were unable to pay their board. Some left Nicholson's employ. Others received help from Pollard. Although he had "not a dollar to go to Market for [his] own Family," Pollard not only allowed his workers to bill their rent to his account to counter the threats of their landlords but permitted them to draw orders on his shoemakers. A network of lending and borrowing also developed among the workmen. But in midsummer, with no wages coming in, Pollard's workmen began suing each other for debt. By the end of the year, Nicholson had dissolved his relationship with Pollard, who could no longer attract workmen because of his "bad credit as a paymaster."[37]

The distinction between manager and worker had blurred as Pollard, along with the day laborers, faced the economic hardships caused by Nicholson's vacillating funds. Although frustrated at times with his workmen's behavior, Pollard was a skilled machinist himself and respected his craftsmen-employees. In the summer of 1795, he exhibited a paternalistic sense of responsibility for them. The workmen stayed with Pollard until his struggles to obtain wages for them proved futile.[38]

The workmen in the button manufactory and their supervisor, Jonathan Mix, adopted a different strategy to ensure their livelihoods. Mix, who had successfully manufactured buttons in New Haven, angrily blamed Nicholson for not honoring the terms of their contract and undermining the success of the business. Nicholson had not kept the button works supplied with tin. Without necessary raw materials, the manufactory, which one English manufacturer estimated could produce an annual profit of £3,000, was "doing nothing."[39]

At the end of February 1796, the twenty-six employees of the button works told their supervisor they would not continue to work unless they were paid the wages due them. They were ready to join Mix at a new factory in Providence. The workmen's accounts indicate that they earned half of the monthly income of the other Nicholson manufactory workers and thus could withstand less easily the delays in their wage payments.[40] The button workers had an additional reason to feel exploited by Nicholson. In the summer of 1795, he was offering books instead of cash for their labor power.

Nicholson had never paid wages strictly in specie. Rather he had arranged for the workers to take up to half their earnings in store goods and rent. As his financial situation worsened, Nicholson relied more and more on the "truck" system. He stocked the Falls store with flour, potatoes, beef, and cider, which he required the workers to take as wages.[41] In June 1795, however, Nicholson had his supervisors pay the button workers' wages in the form of books. While the laborers in the button manufactory were literate men, the price of the novels and histories made them a luxury for any wage earner. Some workmen were debited for *Citizen of the World*, which at the price of a dollar required over two day's labor to purchase. Other volumes, such as *Cook's Voyage* or *Letters of Sir Chesterfield*, cost a day's wages. Some workers were also charged for two or more copies of the same book. Lyman Atwater, for example, was debited for two volumes of *The American Revolution*. Whatever his historical or political tastes, he, like the others, was burdened with selling his books in order to procure his earnings.[42]

The button workers turned on their employer a year later by conspiring with Jonathan Mix to embezzle and sell surreptitiously buttons from Nicholson's manufactory. Sears Hubble, a button worker who earned just under fifteen dollars a month, had received books for payment in 1795. Three copies of the novel *Tisela* were debited to his account. A year later he traveled to Philadelphia and Reading to sell boxes of buttons without Nicholson's knowledge. Inconsistent and unsound wage payments had contributed to the uncertain relationship between employer and employee. When Mix scornfully told Nicholson that "collecting the money from you Sir is worse than Earning of it," he captured one of the peculiar dilemmas of the manufactory wage-labor system.[43]

The highly skilled manufactory workmen formed combinations and acted forcefully to redress the grievances against both Nicholson and his supervisors. Some anxious supervisors anticipated violence when wages were not forthcoming. Thomas Flood, the foundry supervisor, quit in fear for his life in June 1795, angrily telling Nicholson that his workers "became so veary nearvy about their money." In the spring of 1796, William Eichbaum pleaded with Nicholson to settle with the workmen after months of undependable wage payments, "I have no peace among them and the Devil knows what they want, only to [breed] a Riot."[44]

Nonviolent collective opposition by the workmen alarmed Nicholson and his managers. In the summer of 1795, the laborers at the steam engine manufactory elected representatives to visit Nicholson and to present personally their demands for wages due them. Charles Taylor, the supervisor, rushed a letter about the impending visit to Nicholson so he would not "be astonished" when the workers appeared at his door in Philadelphia. The stockingmakers turned to the courts in their struggle with Nicholson and supervisor John Campbell. In April 1795, after a week-long work stoppage, six stocking-makers submitted a petition to Campbell, stating their position and demands. Campbell owed them victuals, wood, wool, soap, and candles. They demanded the back wages due them and prompt weekly payments in the future. When Campbell failed to respond, the workmen "lost all confidence" in their supervisor and hired a lawyer to file a complaint in their names against both Campbell and Nicholson. The grievance displayed the motivation of Nicholson's wage laborers. At issue was not wage rates, hours, or the system of production itself, but the irregularity of the wage payments. Campbell had neglected to pay them weekly wages as stipulated in their work contract. Because "their Subsistence Intirely Depends on Their Labour," the wage earners wanted a guarantee that their earnings would be paid "punctually." Nicholson himself complained that the workers were "not affording me much profit," but conceded to the stocking weavers' lawyer that their wages must and would be paid on time. Yet within months, the Fleecy Hosiery Manufactory closed, as Nicholson and Campbell launched into a legal battle with each other.[45]

Undaunted by his difficulty with Campbell, Nicholson sold off the machinery from the hosiery factory and contracted with John Lithgow, the Scottish stocking loom–maker and his partner, William England, to manufacture coarse hosiery. In 1797 England paid a handful of weavers fifty-eight cents per day to produce coarse stockings. As had happened with Campbell, Nicholson could not keep England supplied with cotton and cash for wages. England's weavers, like those of Campbell, demanded that their wages be paid regularly and threatened to sue. It is unclear whether their demands were met. Whatever the case, the weaving factory closed two months later and Nicholson sent England with two remaining hosiery weavers and looms to carry on production in England's home.

The elements of the conflict between manufactory capitalist and laborer were nowhere illuminated better than in the persistent struggle between the glassmakers and Nicholson. Nicholson's glassblowers were highly skilled artisans who depended on a team of apprentice assistants to prepare the materials and shape the bottles. They worked long and difficult hours together in the searing heat of the coal-burning furnaces. The arduousness and the camaraderie of their labor, coupled with their dependence on their wage for subsistence, underlay the wide range of collective tactics these German-born artisans used against their employer.[46]

The largest sector of Nicholson's labor force, the forty-one workmen of the glass manufactory, produced for a very good market. "It is amazing," their supervisor told Nicholson, "to see the orders that daily come without seeking for them." The glassblowers and pot assistants made over three thousand claret and snuff bottles in a good month. However, within nine months of starting production, the manufacture of bottles faltered. In the summer of 1795, William Eichbaum, the supervisor, pleaded with Nicholson for coal, ash, sand, potash, clay, and workmen's wages, reminding him that "the demand for glass is very great and provided we laid in a more convenient situation it would undoubtedly be very profitable." Strapped for capital as always, Nicholson could do little for Eichbaum.[47]

As the pattern of piecemeal payment of wages and unreliable provisions of materials developed, the glass workers organized. In the fall of 1795, after Nicholson sent out a "trifle" of money for wages, the workmen traveled to Philadelphia to confront him personally. Eichbaum informed Nicholson that members of the Fertner family were demanding the individual work accounts in order to sue and had contacted the German Society of Philadelphia for assistance against their employer. The father, Mathew Fertner, had died in the summer, a few days after sending a plea to Nicholson that he had not the "wherewithal to accomplish my Dyet."[48] The conflict intensified through the winter when Nicholson countered the glass workers' demands for past due wages with a charge that they owed him money he had advanced them. He argued that the glassmakers had done "little or nothing" while the "works were getting ready" so they were in debt to him. When they threatened to quit, he informed them that they "could have their discharge if they each paid him ten pounds plus the balance they owed him."[49] The resolution of the conflict is unclear in the surviving records. Knowing that he could "ill spare them," Nicholson apparently came up with the money to settle their accounts. Whatever the case, the battle illustrated how Nicholson's inability to meet wage payments involved capitalist and laborers in a protracted dispute over the terms of employment.

The difficulties created by this unsystematic and uncertain wage system became evident again in June 1796 when Nicholson attempted to pay the workmen in paper notes after it became known that he held a sufficient amount of specie. Learning of Nicholson's deception, John Sweetman, a glassblower, informed his fellow workmen. Nicholson soon received warnings from the Falls that "the glassmakers here are furious and let me tell you that there is no time to be lost."[50] The glass workers collectively refused to accept their wages in notes and forced Nicholson to meet their demands for payment in hard money.

As long as Nicholson consistently settled the accounts of the workmen, they labored steadily. Whenever wages were not paid on time or in a fair manner, the workers suspended production. It is important to note that the undependable behavior of the glassmakers that prompted frequent complaint among the supervisors—the work breaks and drinking binges—occurred during or soon following the times of irregular wage payments. In the fall of 1795, for example, Eichbaum's protest that his glassmakers had left the manufactory with "glass in the fire" and made no glass coincided with Nicholson's forwarding of only a "trifle" of wages to the Falls. The two days of "Bacchus' Reigne" came on the heels of the wage rate dispute between Nicholson and the glassmakers in December 1795. The interest and attitude of the early manufacturing wage earners will be misinterpreted if we loosely apply what Gutman has labeled "the strange and seemingly useless work habits" that so plagued early manufacturers. From the fall of 1795 through the spring of 1796, the "undisciplined work habits" of the glass workers appear to have been primarily a form of labor protest.[51]

In the spring of 1797, a strike by the glass workers finally forced Nicholson to close the Falls complex. The glassmakers had halted production in March when, once again, wages were not forthcoming. Nicholson was shocked by the losses "sustained by their refusal to work" and implemented the "fifth piece" wage rate. This traditional form of payment-in-kind required the glassblower to take as payment one bottle for every five he produced. The exploitative nature of this system was the same as Nicholson's payment of books to his button workers. Its essense was captured in Nicholson's instructions to Eichbaum: "They may take the fifth piece and turn it into money and pay themselves."[52] But in order to turn one-fifth of the product of their labor "into money," the glass workers, in much the same manner as the buttonmakers, had to travel to Philadelphia to sell their bottles and thus lose days of earnings.

Within weeks the glass manufactory laborers refused to accept the fifth piece as payment. In May they made clear to Nicholson that they were willing to work only if he paid money wages. On June 2, the glass workers gathered in their supervisor's office. When Eichbaum ordered them to prepare the glass that was ready, they informed him that they would "not work a stroke unless they get all the money which is due to them." This standout proved to be the final blow to Nicholson's manufacturing venture. Upon receiving word of the strike, Nicholson ordered his agent at the Falls to discharge the boatmen (who transported the bottles) and the glassmakers and to "let the fire go out at the glass works."[53]

The failure of the manufacturing enterprise at the Falls of Schuylkill was a bitter setback for Nicholson's skilled supervisors. Their fate illuminated the problematic transition of the master to manufacturer in the early years of capitalist manufacturing. Exceedingly enthusiastic about the profit to be made in manufacturing in America, British and German masters joined their skills with Nicholson's capital. As supervisors of production, they endeavored to meet their dual responsibility. They strove on the one hand as partners of Nicholson to maintain profitable production. At the same time, they struggled, to varying degrees, on behalf of their workers and fellow craftsmen to obtain the wages they earned. With Nicholson's failure, they grappled to preserve their identity and livelihoods as masters of their trade.

Although William Pollard, Charles Taylor, John Lithgow, William England, Thomas Flood, and Jonathan Mix had experience and knowledge in management and marketing, they were essentially skilled craftsmen and machinists. Their identity as artisans was manifest in the pride they held in their skills. This was evident, for example, when Flood, the foundry expert, lashed out at Nicholson for disgracing him by his inadequate effort, in Flood's view, to set up and supply the foundry at the Falls complex.[54]

Those manufacturers who severed ties with Nicholson early in the turmoils over construction and control of machinery were the most fortunate. One example was John Bowler, an English inventor of carding and spinning machines with whom Nicholson had contracted to build his cotton machinery. In a dispute with Nicholson over funding, Bowler decided to return to London where he had left his family. He sold the machinery or took the models with him, forcing Nicholson to forfeit $10,000 he had originally invested. Nicholson's loss came at the same time that he was suing Campbell over ownership of the stocking looms he had built.[55] Nicholson lamented to Pollard, "thus you see one after another how my manufactories serve me."[56]

Yet those manufacturers who did serve Nicholson faithfully and competently as supervisors in his manufactories saw their livelihoods sink with his venture. Only William Eichbaum, who found a job in a glass manufactory in Pittsburgh, seemed to be the exception in a string of individually lamentable tales.[57] With Nicholson bankrupt and their respective manufactories closed, Pollard, Taylor, England, and Lithgow turned to their experience as craftsmen and machinists and struggled in a pathetic way to hold onto the machines that had once represented their economic independence.

Charles Taylor, the builder of Nicholson's steam engine and cotton machinery, trusted Nicholson's assurances of continued funding of the steam engine and cloth works. But by the spring of 1797, Taylor was desperately looking for other employment, informing Nicholson that "your present circumstances will not admit of you to answer the absolute necessity or want of my Family."[58] The following year Taylor, who came to Philadelphia with an international reputation as the builder of England's Albion Mills, was still in Nicholson's employ, retained in the simple task of keeping the idle "Cotton Machinary that cost . . . so much money . . . from rusting" and in repair.[59]

The other textile experts endeavored to take up their old trades. John Lithgow turned to "keeping a small grocery store" in Philadelphia after Nicholson closed his enterprise. Yet up until Nicholson's death in debtor's prison in 1800, Lithgow tried to arrange for a loan of a stocking frame to work. His partner, William England, managed to keep possession of his stocking frames despite threats of seizure by the sheriff. Six months after production had stopped at Nicholson's complex, England again implored him to fulfill his promise of capital. But the hosiery manufactory existed in name only, lacking laborers, raw materials, and capital. England reported to Nicholson, nevertheless, of his circumstances:

> I have labored under the great Difficulty this six months past being obliged to make a few pairs of stockings and hawk them about town and for the sake of ready money to sell them at little more than half price. But of late I cannot sell them at all. I have no means of subsistence at this place.[60]

When England died in 1799, his frames were part of his small estate. William Pollard, who had introduced the waterframe to Philadelphia under Nicholson's auspices, also held onto his machine. He died in debt; his demonstration model of the Arkwright frame was the principal asset in his possession, symbolic of his achievements as a manufacturer.[61] The experience of these skilled manufacturers, like that of Nicholson and the laborers, underscored the conflict that accompanied capitalist manufacturing in the new republic.

Although the Nicholson manufacturing venture is only a single case, it offers evidence that the relations of capital and labor developed fitfully in the widening market economy of the 1790s. Nicholson sank too much of his capital into machinery, buildings, and land, leaving little for operating expenses at the Falls. He had also invested in other enterprises that consumed his capital without producing profit. At the Falls complex itself, Nicholson could not coordinate the operations of various branches despite his careful system of accounting.[62]

Equally important to Nicholson's failure as his mismanagement of capital were the discrepancies within the developing factory wage-labor system. Nicholson begrudgingly bought labor power at high cost. He criticized his workmen for behavior in and out of the workplace that countered profitable production. Yet he could not create a sound and consistent system of wage payments. Vacillating between piece rates and daily wage rates, Nicholson manipulated the form of payment in kind and money and settled wage accounts on a piecemeal basis. As wage earners, the workmen contracted to labor at rates to provide their subsistence. They expected wages to be paid regularly in a systematic fashion and condemned Nicholson's inability to meet his obligations. They ignored their responsibilities on the job, took legal action, and stopped work—combining customary and more institutional actions—to force their employer to rationalize wage payments. The conflict between employee and employer hinged on the wage laborer's irreducible need to ensure regular earnings and the inability of the capitalist to provide wages in exchange for labor.

The nonmechanized manufactory of turn-of-the-century Philadelphia represented a transitional arena of production in which skilled workmen, mechanic-manufacturers, and merchant capitalists struggled to realize their interests as manufactory wage-laborers, managers, and capitalist manufacturers. Nicholson, his managers, and his laborers possessed a mixed set of interests and abilities that complicated their roles in the nascent factory wage-labor system and hampered the starting up and operation of the Falls of Schuylkill enterprise. While Nicholson may not have typified manufacturers of turn-of-the-century Philadelphia because he overextended and mismanaged his investments, both his perception of labor and the attitudes and sensibilities of those who came in his employ were characteristic of this period.

By observing how customary practices and roles impinged on the relations of production, it is possible to refine our interpretation of the transition from handicraft to industrial forms of production. First, the initial generation of factory workers did not necessarily lose control over the labor process, nor did they perceive themselves as confronting an unfamiliar and alienating wage-labor system. Second, no clear distinction separated the position of manager from owner. Those who supervised production shared the skills of the workmen as well as the profits of the business as ownership partners. Finally, the capitalist manufacturer oversaw management and intervened in disciplining the labor force. Part of the long, uneven, and problematic process of industrialization in the Mid-Atlantic would involve the sharpening of the blurred lines of interest and control between laborer, manager, and capitalist.

It is in the developing Philadelphia cotton textile industry in general that we can follow this process. Hence it is time to step outside Nicholson's world of production and examine the broader forces that were molding the passage of worker and manufacturer into industrial laborer and capitalist.

II

Textiles and the Urban Laborer, 1787–1820

The fitful four-year struggle and ultimate failure of John Nicholson's manufacturing venture was but an early though revealing chapter in Philadelphia's complex transition to industrial capitalism. This microcosm of early manufacturing production has revealed that the supply of laborers, their attitudes as skilled workers, and their needs as wage laborers shaped relations within the centralized workshop. Over the course of the early national period, these factors impinged upon the long-range transition from manufactory production to mechanized factory production.

In Philadelphia, as in England and New England, cotton textiles launched "the industrial revolution." But unlike in New England, machines did not abruptly change the world of work. The manufacture of yarn and cloth went through successive and overlapping stages of handicraft, manufactory, and factory production between 1787 and 1837. A brief outline of the development of Philadelphia textile manufacturing (the spinning of yarn and weaving of cloth) in the era preceding the mechanized mill points up significant differences from New England. First, urban textile manufacturing was based on the labor of the poor, and manufactories were fashioned after English and colonial workhouses. Second, textile manufacturing was moved into mechanized factories relatively late, allowing the traditional labor process of spinning and weaving and the system of outwork production to persist. Last, skilled mule spinners and handloom weavers made Philadelphia into a center of fine yarn and cloth production.[1] A key to understanding the early industrial history of Philadelphia was the city's position as an immigrant port, a status not shared by New England's seaports. Throughout the prefactory period Philadelphia received a steady and abundant supply of Irish and English skilled spinners and weavers, who brought with them a particular set of skills and attitudes that shaped the early course of textile manufacture.[2]

The profitable manufacture of textiles based on the labor of the urban poor had its native roots in the workhouses and linen factories of the colonial port cities. Mid-eighteenth-century manufacturers attempted to ease the tax burden in years of economic distress and, at the same time, to compete profitably with English cloth imports. Both the privately and publicly operated workhouses collected their labor force in a single room or building to spin and weave on company wheels and handlooms. As early as 1739, poor residents of the Philadelphia workhouse spun flax and wove cordage, bagging, and linen cloth for the domestic market. By mid-century, the United Society for Manufacturers in Boston had established a linen yarn factory to employ the growing number of impoverished war widows and their children. While the optimistic managers believed that the system of collecting and supervising poor spinners would be both charitable and potentially profitable, this "first attempt in female factory labor" failed. They could not produce and market linen as cheaply as it could be imported. This was due, in part, Gary B. Nash has argued, to the unwillingness of impoverished women to labor for less than subsistence wages in a factory structure inimical to their traditional work experience.[3]

Parallel fruitless experiments in factory production in the three major port cities during the depression-ridden 1760s show that the lesson was hard learned.[4] The first successful textile manufacturing establishment in Philadelphia, begun in 1775, did, however, adopt a different method of organizing labor. Like its predecessors, the United Company of Philadelphia for Promoting American Manufacture was conceived under the dual notion that making laborers out of the unemployed poor was profitable as well as "providential" work. Rather than collecting spinning wheels and resistant spinners under the factory roof, however, the managers organized a large-scale putting-out system. Hundreds of poor women spun the company's flax and wool at home and, "with great pleasure," one observer noted, returned the yarn for small piece payments. So great was the demand for yarn of all types that the managers of the company advertised to reach every good spinner "however remote from the factory," who would exchange yarn spun from her own materials for ready money. This first apparent success story of manufacturing based on "giving employment to the poor" ended abruptly when the British occupied Philadelphia two years later.[5]

After a decade of war and nation building, Philadelphia entrepreneurs once again set the poor to work manufacturing textiles. In a meeting of the Friends of American Manufacturers, convened in 1787 to establish the Pennsylvania Society for the Encouragement of Manufactures and the Useful Arts, Tench Coxe addressed his fellow Philadelphians on the need for such an institution. At the time when Philadelphia was just beginning to emerge from a debilitating depression, Coxe emphasized how poverty threatened a free government, "rendering people fit instruments for the dangerous purposes of ambitious men." Impoverished men and women could only be made virtuous and obedient by remedying their poverty and making them industrious. Drawing evidence from the remarkable European factories that employed hundreds of women and children, Coxe admonished the gathering that "the employment of the poor in manufactures is of the utmost consequence."[6] In the fall of that year, the manufacturing committee of the society, which included Coxe and John Nicholson, established a textile manufactory with the private subscriptions of "patriotic citizens."

The shareholders appointed a suitable manager for their factory, Samuel Wetherill, a man of "stern republicanism" and an experienced manufacturer of cloth for the Continental army. Wetherill shared the founders' belief that manufactories could be "great exertions of virtue and industry" as they opened "extensive fields of employment for persons of almost every description." A successful Quaker entrepreneur, Wetherill also intended that the production of yarn and cloth would prove "a source both of private and public wealth."[7]

The manufacturing committee started operations with a labor force of female wheel spinners. In order to manufacture linen yarn successfully, Wetherill organized them into a putting-out network, a system of production that British textile merchants had found profitable for most of the eighteenth century. The labor of the domestic outworker—in this case a female flax spinner—was in theory cheap and dispensable. As an experienced textile manufacturer familiar with competition from English imports and periodic market downturns, Wetherill must have found at least two advantages in the domestic putting-out system: capital was not immobilized in buildings and equipment, and labor was easily laid off.[8] The managers, of course, promised that the domestic system of production would "not only be advantageous to the society, but give employment to many of the industrious poor." On market days during the winter and spring of 1788, a superintendent gave out flax to hundreds of urban women, who applied with "recommendations of their honesty and sobriety in hand."[9] Although the committee boasted that providing flax for spinning would enable the industrious female poor of the city to support themselves and "become useful members of the community," the scanty wage accounts suggest differently. Spinners did not produce yarn regularly or earn a steady income in the manufacturing society's putting-out system.

It was not, of course, this attempt at a large-scale outwork system for the city's poor women that distinguished the PSEMUA from earlier urban textile establishments. Production within the society's factory on the corner of Ninth and Market Streets marked two important changes in manufacturing: the introduction of mechanized spinning and the collection of handloom weavers in centralized production.

Philadelphians were treated to a glimpse of the new machines of the society in the grand Federal Procession of July 4, 1788. The proudly displayed brigade of the PSEMUA encapsulated the early world of mechanized cotton textile production. On a thirty-foot carriage covered with cotton rode the devices and their attendants from the society's Market Street factory. Two people operated a carding engine that could process fifty pounds of cotton per day; one woman worked a spinning jenny that contained eighty spindles. Also on display was a large handloom operated by a single weaver using a fly shuttle and a cloth-printing machine. Behind the carriage full of spinning and carding devices strode the handloom weavers of the factory joined by other members of this large urban trade. The city's wheel spinners, whose livelihoods were threatened by the manufacturing society's new jennies, did not march in the procession.[10]

The managers had set up twenty-six handlooms in the cellar of the "old factory" building on Market Street in the fall of 1787. They gained greater control over the regularity, speed, and quality of production by collecting a number of traditionally independent weavers under one roof. This system also reduced the likelihood of embezzlement, a practice of outwork weavers that had long tormented British weaving masters. A New England manufacturer in 1809 noted the great advantage of the centralized form of cloth production: "One hundred loom weavers in families will not weave so much cloth as thirty, at least, constantly employed under the immediate inspection of a workman."[11] One employee of the weavers' branch at the Market Street factory supervised the laborers and was empowered "to discharge such workmen as do not complete their work in proper manner and agreeable to his direction."[12]

This early factory weaver did not own or rent his loom, control the work pace, or own the product of his labor, yet the labor process itself had not been divided or mechanized. The manufacturing society's weavers, like John Nicholson's glassblowers and frame builders, thus retained something very tangible as manufactory laborers—their skill and earning power. Some did choose to rent looms from the society when they had "sufficient security" to do so. But the several poor weavers employed at the factory commanded the same piece rates as their independent counterparts. It was with a sense of commonality that the factory weavers marched with others of the same trade behind the manufactory's carriage in the Federal Procession.[13]

Above the weavers' branch in the factory's cellar, however, the PSEMUA installed machines that significantly altered the labor and livelihoods of the female wheel spinners. The spinning jenny, which was so proudly displayed on the parade carriage, was one of four erected in the "long room and garret" for the manufacture of cotton yarn in early 1788. It not only increased the productivity of a spinner many times over, but, like the centralization of weaving, gave the managers greater control over the process of production.[14]

Patented by a Lancashire weaver in 1770, the spinning jenny simply multiplied the number of spindles one spinner could work.[15] The first jennies used only eight or twelve spindles and women and children operated them. In 1787, however, the manufacturing society constructed a jenny that replicated the hand process of stretching and winding on eighty spindles. Once these machines contained over a dozen spindles, the carriage was too heavy for women or children to move. Indeed, by the time Philadelphia had its first glimpse of this labor-saving device, men had displaced women in the spinning industry in England, since female hand spinners simply did not have the physical strength to operate these machines. These "multiplying wheels," by reducing the cost of yarn, caused widespread distress among the female hand spinners in both the woolen and cotton districts of Britain.[16] The four jennies that the manufacturing society operated on its factory floor contained forty to eighty spindles each and probably were operated by a man with the assistance of a child who pieced broken threads together. This device not only threatened the demand for the Philadelphia female spinner's linen yarn but reduced the price of her handspun product of similar quality by 25 percent.[17] It was for these reasons, perhaps, that the two to three hundred hand spinners employed by the PSEMUA did not join the brigade of male weavers in celebration of the city's first cotton textile manufactory.

Local demand for the factory's cloth products proved to be vigorous. Philadelphia cloth merchants negotiated to have their weaving done by the Market Street factory. Expanding sales and production compelled the manufacturing committee to subdivide the management of the business of the factory and to contract with the merchant firm of Mendenhall and Cope to sell their manufactured goods. Workers were in plentiful supply. So many of the port city's unemployed weavers crowded into the factory for jobs that production was obstructed, and the managers agreed thereafter to hire only those recommended by company shareholders. The manufacturing committee even sought to rent, though unsuccessfully, a large wing of the city's House of Employment for more production space.[18]

Within six months, the committee was trying to attract to the city English mechanics who could build the Arkwright perpetual spinning frame. Given that their successful trial with the spinning jenny had proved the "practicability of establishing the cotton manufacture in this city," the PSEMUA intended to tap the extensive home market for coarse-spun cotton. In the same month that the committee penned a notice to entice machine builders to Philadelphia, the state government purchased one hundred shares of the society's "manufactory of cotton articles," noting it had "been established with great prospect of success."[19]

Yet the manufacturing society never had the opportunity to test the Arkwright waterframe or the Philadelphia market for cotton cloth. In March 1790, a fire destroyed the combined spinning and weaving factory of the PSEMUA along with materials and goods worth £1,000. The shareholders had received threatening letters and were convinced that the factory had been burned by design. They offered a reward for the capture of "the persons who willfully and maliciously" set fire to the factory.[20]

It seems likely, particularly in light of the inglorious introduction of the jenny to England, that Philadelphia wheel spinners whose piece rates had been reduced became machine breakers and destroyed the devices that threatened their livelihoods. In the early 1780s, machine breakers demolished all the jennies with more than twenty spindles—"the hated rival of their fingers"—in towns throughout Lancashire. They selectively left intact those that were small enough to be operated by women and used in a cottage. Even the original machine was destroyed and its inventor, James Hargreaves, driven out of the district by neighboring yarn producers.[21] Similarly, the Mid-Atlantic's first attempt at mechanized textile production had been cut short, it appears, by the violent action of urban textile producers. Although there is little concrete evidence of machine destruction between 1790 and the period of widespread mechanization in the 1820s, the Friends of Manufacture were fearful of the possibility of industrial sabotage. In 1804 Tench Coxe, citing the destruction of the PSEMUA factory on Market Street, warned the urban community of "the danger of casual and intentional conflagrations in the Labour-saving manufactories."[22]

Although Philadelphia's first cotton spinning and weaving factory was short lived, its brief history suggests a course of development that varied from the rural New England textile enterprises that would follow. The manufacturing committee of the society moved to centralize both handloom weaving and the spinning of coarse cotton and at the same time maintained a large outwork system of female flax and wool spinners. Their labor force of male weavers and female spinners was drawn from the ranks of the city's laboring poor. Underemployment plagued the ranks of both urban trades. Jobless weavers crowded into the society's factory and were turned away. The spinners' need to protect their livelihood caused their turn to incendiarism.

The material position of the city's weavers and spinners and the problem of urban poverty would continue to shape the way textile capitalists organized labor and production. In 1790, the year that machine breakers apparently destroyed the first "labor-saving manufactory," the public campaign to mechanize spinning took off. The seaport's labor supply was central in this Philadelphia-based movement to promote the Arkwright spinning frame and factory system.

Every promoter of manufacture made explicit the need to apply the English spinning devices to production because of the coexisting dearness of skilled labor and abundance of unskilled laboring poor. "In a Country like America," one Scottish observer pointed out to Tench Coxe, "they [labor-saving machines] must be Doubly advantageous where the Labour is so high."[23] In the late 1790s, the first individual to establish a successful water-powered spinning mill in Philadelphia confirmed the feasibility of the Arkwright spinning frame for answering the particular needs of the city: "Where manual labor is so expensive as in this country, the adoption of a machine to work with such facility and accuracy, must be an object worthy of general encouragement."[24]

The benefit of the British spinning frame did not need to be pointed out to Alexander Hamilton or Tench Coxe, the most ardent and politically powerful promoters of textile manufacturing, who took it upon themselves to develop the argument for linking female and child labor to the labor-saving machinery. Besides the lack of capital, the "dearness of labour" was the greatest impediment to the success of manufacturers, Hamilton pointed out in his 1791 prospectus for the PSEMUA. He went on to note that recent improvements in technology had decreased "the proportion of manual labor" and speculated that "women and children in the populous parts of the country may be rendered auxiliary to undertakings of this nature."[25] The fact that America could manufacture without taking people from agriculture was a much emphasized point.[26] "It will not take off one hand from the plough," one advocate of manufacturing assured Coxe. While the employment of this "new and large class of persons" in manufactures would preserve the agrarian backbone of the nation, it had a more practical advantage: females were a significantly cheaper form of labor. Employing women, who would "work for low wages," in the manufactories would "prevent the rise of wages among the male manufacturers."[27]

English factories at this stage were an enticing prototype. Their water-powered mills with "machines ingeniously contrived" employed "a few hundreds of women and children, and perform the work of twelve thousand carders, spinners, and winders," Coxe marvelled. Hamilton wrote in awe that, "attended chiefly by women and children," the cotton mills operated "during the night as well as through the day."[28]

The promotion of textile factories was further enhanced by paternalistic republican precepts that legitimized putting poor females and children to work in the mills. Hamilton felt it worthy of particular remark that "women and children were rendered more useful . . . by manufacturing establishments than they would otherwise be."[29] The potential operators of the Beverly Cotton Factory in Massachusetts petitioned the state that their factory would "afford employment to a great number of women and children, many of whom will be otherwise useless, if not burdensome to society." The advocate who wrote as "Juriscola" believed it was the duty of patriarchal policymakers to bring females into textile manufactories. "Justice, policy, and benevolence ought to excite us to better the condition of the female sex whenever we can." The premise and conclusion were straightforward: "Every city man is taught a trade or calling; every country man is taught the same or to plough, harrow, sow, and thresh. Every city and country woman should be taught to card, spin, weave, and dye."[30] There seemed to be no objection in the aftermath of the postwar depression to placing women and children as operators of the labor-saving spinning mills. Output would escalate, the cost of labor would fall, and a large sector of the city's poor would be made useful and industrious.

Yet in the last decade of the eighteenth century, Philadelphia yarn production followed a quite different course than that envisioned by the republican promoters of manufacturing. As the Arkwright machines and the water-powered cotton mills that housed them quickly took root in New England soil, the traditional modes and organization of spinning persisted in the Mid-Atlantic seaport. The city woman was a wheel spinner, not a machine attendant. As her ranks expanded, so did that of the male handloom weaver and mule spinner.

In explaining why yarn as well as cloth production thrived in an urban putting-out system, it is important to examine the position of the traditional textile producers—the female wheel spinners and male handloom weavers and mule spinners—who lived and labored in the port city. In the postrevolutionary period, thousands of displaced Irish and English textile workers disembarked in the seaport seeking to carry on their trade. Their numbers, their low cost, and their skills impeded the entrepreneurial impulse to mechanize spinning and weaving through the early decades of the nineteenth century. In contrast, the New England ports, around which the labor processes and productive relations changed most rapidly, received few textile immigrants. The first New England spinning mills relied instead on children from the surrounding countryside.

During the three decades before 1812, when America and England were not at war, a continuous immigration of spinners and weavers from England and Ireland fed the populations of the Mid-Atlantic port cities. Between 1783 and 1789, over 24,000 Scotch and Irish arrived in ports along the Delaware River. Over the next four years, Philadelphia received an annual average of 2,943 immigrants, over half of whom were from Ireland. In these last two decades of the century, approximately 5,000 Ulster passengers per year left for America, and the majority of the ships that set sail from Londonderry, Belfast, and Newry took their passengers to Philadelphia and to Wilmington and New Castle, Delaware.[31]

The mechanization of cotton spinning and the decline of prices for handloom woven products were the general developments that compelled linen and wool wheel spinners and weavers to leave Britain. Many of the passengers departing Ulster and southern ports in Ireland had been employed in the overcrowded and depression-ridden linen industry.[32] The introduction of the Arkwright water-powered frame had virtually eliminated the livelihood of the female hand spinner throughout Britain before the end of the century. This machine, which, unlike the jenny, completely mechanized the spinning process, produced a tough cotton yarn suitable for warp.[33] As the coarse, cheap, mill-spun yarn from Lancashire flooded the markets in the 1780s, flax spinners from Ireland and wool spinners from the southwest of England lost their market for warp. The position of the British independent jenny spinner had also deteriorated by the turn of the century as most fine yarn production had entered spinning houses containing large multispindled jennies. The growing demand for handloom weavers to work up the great bulk of cotton yarn brought prosperity to that trade only until the onset of war and depression in 1793.[34] In the early 1790s, eager Philadelphia manufacturers were anticipating the immigration of Britain's "unfortunate competitors" looking for employment in America.[35]

Indeed, faced with pauperization at home, many textile workers boarded ships bound for America. "Guineamen with slaves were never crowded like the American ships from Londonderry to Philadelphia with Irish passengers," reported a turn-of-the-century visitor to the city.[36] Eyewitnesses in both British and American ports identified a large number of the passengers as poor weavers and their families. "There were not any people of real property," an Irish customs officer at Newry reported, among the weavers, smiths, and joiners that set sail on three vessels for American ports in 1783. A contemporary American correspondent confirmed that it was the "lower order of tradesmen" who were heading across the Atlantic—"the industrious, careful linen weaver, who has scraped together a sufficiency to transport himself and family." In 1794 the traveler Henry Wansey observed two hundred Irish weavers leaving one ship in the port of New York. Another Philadelphia tourist reported that "fourteen thousand souls were landed from Ireland by the Philadelphia ships alone in 1801."[37]

Until the restrictive Passenger Act of 1803, a linen weaver or spinner could board a ship at Londonderry bound for Philadelphia for about £3.10s. Many of those who did scrape together the relatively small amount for passage did not find their way to hinterland farms but remained in the port where they disembarked. Weavers found themselves competing in a glutted labor market. For example, William Byrnes, a weaver by trade, arrived in Philadelphia in the summer of 1792. He survived just two months in the city before entering the almshouse.[38]

In 1790, Philadelphia's established Irish-born community had organized a Hibernian Society to assist Irish emigrants "whom misery, misfortune, and oppression had compelled to forsake their native country." Those who landed without friends in the city "suffer considerable at first," Matthew Carey, a founder of the society, wrote of the three to four thousand Irish that arrived in 1791. By the end of the decade, the extreme poverty of the Irish immigrants had turned them into an urban underclass.[39]

Mechanization of New England's cotton yarn industry further exacerbated the position of Philadelphia's female yarn producers. In 1791, Samuel Slater duplicated the Arkwright spinning frame and joined with Providence merchants William Almy and Moses Brown to establish the first cotton mill on American shores. The continuous spinning machine radically altered the labor process of spinning so that the mills virtually needed no more than a work force of unskilled rural children. The only adults employed in the Almy and Brown mill were the overseers and repair mechanics.[40] By the time of the Embargo of 1807, over a dozen of these cotton mills operated in Rhode Island and Massachusetts.

Philadelphia hand spinners, like the spinners in the southwest of England, felt the competition of these distant factories as the price of their labor was reduced. The water frame necessarily produced a coarse yarn, since the cotton rovings had to withstand the doubling and pulling performed in the uninterrupted process of drawing, twisting, and spinning. The cheap coarse cotton yarn was suitable only for warp and competed directly with the hand-spun traditional linen warp. Philadelphia weavers and merchants became major processors and distributors of this New England cotton warp. Almy and Brown wholesaled 16 percent of their yarn through their Philadelphia agent in 1806 and 30 percent in 1808. After decades of stable prices, Philadelphia linen yarn spinners experienced a decline in the piece rates for their linen product when the machine-spun cotton entered the Mid-Atlantic market.[41]

Few attempts were made in Philadelphia before the Embargo of 1807 to build the machines and mills to manufacture cheap cotton yarn.[42] Philadelphia textile capital flowed not only into handloom weaving but into fine cotton mule spinning. A sizable population of mule spinners fed by the migration from Britain turned the port city into a center of fine as well as coarse yarn production. In 1804, when Almy and Brown advertised for a man who knew how to spin on "a machine called a mule," they received several applications from Philadelphia.[43] Developed in the 1780s, the mule was a complex machine that combined the drawing process of the water-powered frame and the stretching operation of the jenny. The machine required a skilled spinner who could operate the carriage with one hand, while managing, with the proper amount of tension, the fly and spindle that wound the yarn. The mule spinner, invariably a male, produced a yarn that was as fine as that spun by the jenny spinner yet as strong as that produced on the water frame. The New England coarse cotton yarn could not compete with the product of the Philadelphia mule spinner, nor could the middle-range yarn of any Philadelphia spinning jenny.[44]

The statistics that Tench Coxe compiled through the first federal census of manufactures illuminates the configurations of Philadelphia's'a textile labor force and the modes of yarn and cloth production in 1810. The numbers alone—of machines, spinners, and weavers—are telling evidence of how different the world of production of the Mid-Atlantic seaport was from that of New England.

Sixty-one cotton mills, with 31,000 spindles on water frames or throstles, were in operation in the country as a whole in 1810. Over one-quarter of these cotton spinning factories (seventeen) were large mills in Rhode Island, which operated almost half of all the spindles in the United States. Philadelphia city and county had one-third the number of spindles on labor-saving machines of the Rhode Island mills, and most of these machines were mules or jennies. When Coxe counted the labor-saving machinery in the city and county of Philadelphia, he found seventeen mules, twenty-one jennies, and a handful of water-powered frames. Such a substantial share of the port city's yarn was produced by hand spinners that Coxe enumerated the number of spinning wheels. He found 3,648 traditional wheels in the city and county, indicating that more females produced yarn in Philadelphia than the total number of women and children reportedly employed in the nation's cotton-spinning factories.[45]

The 1810 manufacturing census also identified Philadelphia as the center of handloom weaving. Two hundred seventy-three handlooms were enumerated in Philadelphia, which equaled a quarter of the looms in the entire state of Rhode Island. Philadelphia weavers used fly shuttles to produce mostly cotton cloth, as well as to process the linen and wool yarn from the local spinners. They not only duplicated the Rhode Island weavers in the production of the coarse cotton goods—bed ticking, shirtings, and sheetings—but competed with England in the production of muslins and other fine cloths.

In contrast to the Rhode Island weavers, many of them rural women, Philadelphia handloom weavers labored full time at their urban trade. Most of the port city's weavers probably produced cloth within their homes or small shops for merchant weavers who controlled putting-out networks. Coxe reported eight cotton cloth manufacturing establishments of which two were in the city; yet we have little direct evidence on the size or nature of these centralized shops.[46] However, we can turn to another source—the accounts of the manufacturing committee of the Guardians of the Poor—to discern more clearly the nature of textile manufacturing and the social factors that inhibited mechanization and perpetuated the traditional forms of worklife of female and male textile producers.

In 1807, as was the case in 1787, the capitalist manufacture of textiles depended on the organization of poor weavers and spinners into domestic networks and, later, centralized workshops. In 1806, for example, a group of entrepreneurs had drawn up plans to operate a brokerage house for the products of "unfortunate" weavers and spinners of the city. Described as "patriotic individuals," they intended not only to obtain "profits or interest on the capital" invested by the subscribers but to "encourage and stimulate the industry of persons of small means." They anticipated that both domestic weavers and hand spinners would deposit their cloth and yarn in the large textile warehouses.[47] It was no anomaly, then, that the Guardians of the Poor, who heretofore had concentrated on maintaining the dependent poor, became the largest employers of textile laborers before the War of 1812. Besides being a direct descendant of the United Company of Phila-

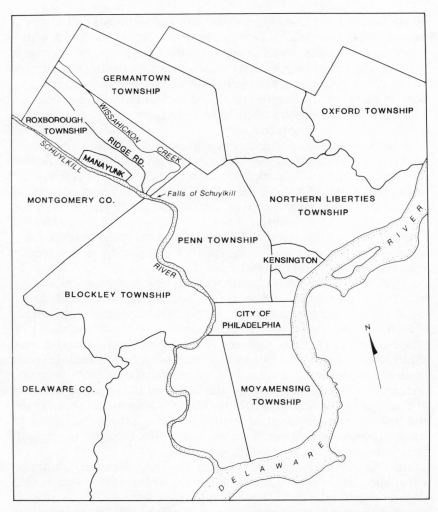

The textile districts of early nineteenth-century Philadelphia (adapted by Matt McGrath from John Daley and Allen Weinberg, "Geneaology of Philadelphia County Subdivisions," City and County Archives of Philadelphia, 1966).

delphia and the PSEMUA, the move of the Guardians of the Poor into capitalist manufacturing fit the mold of eighteenth-century English workhouse manufactories, which were designed to derive "profit from the labour of the inmates."[48] There were few areas in England, Sidney Pollard has argued, in which major textile work was not associated with workhouses and orphanages.[49] Thus the manufacture of textiles by the city's overseers of the poor should be considered historically peculiar only if we consider the New England cotton mill as the paradigm of early industrialization.

The Daily Occurrence Dockets of the city's almshouse as early as 1790 indicate that the Guardians of the Poor were employing female inmates on the spinning wheels of the poorhouse. In the first few years of the new century, however, when French and American trade blockades curtailed British cloth exports, the tax-supported Guardians developed a textile manufacturing enterprise. They gave control to a manufacturing committee who organized and managed the domestic production of flax and woolen yarn and the factory production of cotton yarn and all types of cloth.

Appointed to direct the textile operations of the almshouse was a mix of experienced entrepreneurs and professionals, men such as David C. Claypoole, editor and publisher, and the merchant Joseph Paul. They carefully calculated production to the shifts in the marketplace and the availability of materials. While the small Rhode Island cotton mills concentrated entirely on carding and coarse-spinning cotton yarns to sell to Boston merchants and Philadelphia merchant weavers, the Guardians of the Poor tapped the changing demands for a whole range of cotton, linen, and woolen cloths as well as tow yarn and thread.[50] When there was a shift in prices or demand, the factory steward received a directive from the manufacturing committee to switch the weavers from one type of cloth to another, to adapt the machines to a different quality of yarn, or to hire only those weavers who could produce a specific kind of material. The committee also kept abreast of the changing quality and prices of the raw materials, turning to New Jersey for reasonably priced flax, purchasing some East India cotton yarn in the early years, and shifting from West India to upland Georgia and then to Sea Island cotton as their machinery and market demands changed. Within the different departments of the factory, all the steps of cloth production, from the hackling of flax or picking of cotton to the bleaching of the woven linen or sheeting, were carried on within a form of organization that would not be introduced in the mechanized New England mills until the adoption of the powerloom during the War of 1812.[51]

The Guardians of the Poor put both resident paupers and "outdoor poor people recommended by the Guardians' orders" to work hackling flax, weaving stockings and cloth, attending carding and cotton-spinning machinery, and spinning on wheels and mules in the factory and the home. In the early years, the committee had a ready supply of both indoor and outdoor spinners and weavers among the city's poor. In February 1807, for example, the committee on manufactures directed the factory superintendent to "set up Hatchels in the flax loft" after finding "a Number of Paupers acquainted with the Hatchling of Flax" in residence. Three to four of these male inmates of the almshouse labored year round preparing the fiber of the flax for the spinning process, earning one cent for every pound of flax they processed. In September 1808, the committee began the production of stockings in the factory after considering the "expediency of erecting a stocking loom and employing such of the paupers as are acquainted with the business."[52]

The textile manufacturers also put a number of the young pauper boys to work attending the carding machines and assisting the jenny or mule spinners. The Guardians normally bound out the resident children when they reached the age of fourteen. As the boys' labor became valuable in the working and carding of cotton, however, the committee resolved to hold onto them, binding them as apprentices to the manager to learn the trade of carding and spinning.

The committee hired indoor weavers and mule spinners to supplement their factory labor force of unpaid indoor paupers.[53] Two or three skilled employees operated the factory mules to remedy the "deficiency of cotton yarn in the manufactory" in the last months of 1807. The Guardians of the Poor installed the improved mules to increase productivity. In the first few years of the factory's operation, a combination of over thirty indoor and outdoor weavers found year-round employment producing linen and cotton cloth to be sold by the manufacturing committee. The superintendent put the dependent poor—the resident paupers—to work on the factory cellar looms, paid wages to tradesmen who boarded at the almshouse and worked the company looms, and hired weavers who labored within their homes at piece rates.[54]

Far more female "outdoor" wheel spinners worked for the Guardians of the Poor than any other group of tradespeople. Two hundred and fifty-six female spinners appeared on the account books in 1806–1807 and almost half had Irish surnames. Between November and March, the peak production months, over two hundred women made, on the average, five or six trips to the almshouse factory to pick up mostly flax and to return the tow and linen yarn they had spun on their own wheels. The factory also put unpaid female paupers to work spinning on wheels under the supervision of a matron. Until 1810 these "indoor" spinners numbered between forty and sixty, far less than the number of spinners who owned their own machines and labored in the putting-out system.

The Guardians of the Poor operated its textile manufactory through the second decade of the century yet never adopted the labor-saving water frame. In the months following the December 1807 embargo, the demand for American-produced textiles soared. The Guardians' manufacturing committee immediately drew up plans to expand production. The labor force of indoor and outdoor wheel spinners could not keep the looms supplied with enough yarn to meet the rapidly expanding market.[55] The minutes of January 13, 1808, report that the steward of the factory warned the committee that the factory's present "cotton spinning was an old and contracted plan" in comparison to most other states. He recommended that the "new and improved machines"—the water frames—would be the answer to saving "much time and labor." The committee reviewed a plan to install sixteen water frames and twenty carding machines and additional mules that would employ 280 resident paupers. Their decision not to mechanize the factory, however, soon proved fortuitous. Within four months of peaking, between November 1808 and September 1809, the number of textile workers plunged to an all-time low. The Guardians of the Poor had decided that the shifting ground of the pre-1812 domestic textile market was most profitably negotiated with abundant and expendable labor rather than with the encumbrance of expensive machines.[56]

The social relations of production in the almshouse were modified within a structure of social control in which the employer and employee occupied unusual positions of control and subordination. Since the capitalist manufacture of textiles in Philadelphia had been based traditionally on the labor of the city's dependent and laboring poor, the almshouse factory was an intrinsic part of the world of textile production as the factory replaced the putting-out process. The minutes of the manufacturing committee offer an unusual opportunity to observe the interests and experience of the poorest sector of urban tradespeople and their customary public overseers as they confronted each other as wage earner and capitalist employer.

To promote profitable production, the manufacturing committee carefully governed the activities of their labor force. By January 1808, they had drafted rules and regulations controlling production that stipulated a full-time work schedule. All employees, whether paupers or hired labor, were required to work thirteen- or fourteen-hour days—from sunrise to eight in the winter or to a half hour before sunset in the summer. Supervision consisted of a matron who oversaw the female laborers, a steward who had charge of raw materials, and a superintendent of weavers who coordinated the various stages of cloth production, including spinning. He delivered the flax, cotton, and wool for processing to the hacklers, pickers, and carders. The superintendent then turned over the materials to the steward of the cotton spinners and the matron to distribute to the mule spinners and indoor and outdoor wheel spinners. When he received the spun yarn, the superintendent transferred it to the weavers he personally oversaw. His final responsiblity was to send the cloth to the bleaching green for the last preparation before marketing. Reliable weavers and mule spinners were placed in positions of overseers of their respective departments to assist the superintendent.[57]

The Guardians' attitude toward labor was influenced by their attitude toward the poor. Because they held an idealized notion of how their workers should behave, grounded in the concept of the industrious poor, their management of labor was imbued with a paternalistic attempt to control and shape the values of their employees. Underlying their conduct as managers was the axiom that order, in both individual work habits and the workplace, aided regular production. In January 1810 they noted that production was "in a prosperous way" when the "different manufactories are in good order and the different mechanics orderly." In the minutes of May 15, 1811, the manufacturing committee expressed particular concern that the "weavers' cellars [should be] cleaned and kept in good order that the credit of the factory may be kept up in its usual estimation." The weaving branch of the factory was, of course, one of the most important units of production.

As the Guardians understood it, order and industry began with the character of the worker. They sought out and rewarded monetarily those weavers or spinners who were "of approved character" and labored "in a workmanlike manner." The committee, on the other hand, was quick to punish those who did not follow the precept that "virtue and industry are the springs of happiness." "Indolence," "bad conduct," and "ill behavior" often called for correction. The idle pauper might lose the privilege of eating at the tradesmen's table, while the unindustrious hired weaver faced dismissal.[58]

The manufacturing committee strictly prescribed the daily lives of the factory boys. They either attended the carding and spinning machinery or the classroom. Republican promoters of domestic manufacture at this time believed it the duty of the employer to provide some education for children who labored in the factories.[59] The Guardians of the Poor organized the pauper boys into alternating groups—one of which would "always be employed in the Factory whilst the remainder are receiving the benefit of the school." They placed a bell in "a proper place" to call the children to work and posted the school rules in a conspicuous place in the factory.[60]

Composed of the leading members of the Guardians of the Poor, the manufacturing committee was in a peculiar position to mold the behavior and values of those they put to work within the factory. More important, as employers of the poor, the committee held an unusual share of control over the process of production, labor, and the price of that labor. Once an impoverished weaver or spinner entered the almshouse and became a resident pauper, he or she lost command over his or her own labor. An indoor spinner, for example, by virtue of her dependency, was a bound laborer. She received no payment in exchange for her work, except for overwork (production over the quota) at the piece rate of two cents per pound. The committee set quotas as incentives for their indoor labor force and as measures of their productivity. Both costs and the availability of outdoor workers affected the level of such quotas. At the end of 1808, for example, when the number of outdoor spinners quadrupled from 53 to 200, the committee lowered the overwork level for indoor spinners. On June 14, 1809, the committee directed the steward to furnish them with a list of weavers who did not earn $1.75 per week, presuming they were "absenting themselves from the manufactory unnecessarily."

If boarding costs increased, the indoor spinners were required to produce more. In early March 1811, the committee ordered the matron of the female laborers to enlarge the weekly production requirements, claiming that "as the price of provision, fuel, etc. are greatly enhanced this season, it is good cause to require more labour from paupers supported by the public." Pauper laborers, regardless of their learned skill, had little control over the type of job they held. The superintendent often shifted individual weavers and mule spinners to unskilled tasks.[61]

Market forces more directly ordered the relations between the textile manufacturers and the outdoor spinners and weavers they hired. These laboring poor remained in control of the production process and, depending on the market, bargained over the price of their labor. At the same time, the manufacturers hired and fired these independent poor as the textile market conditions necessitated. The Guardians of the Poor paid the hired weavers piece rates that were comparable to those received by independent handloom weavers or those in private manufactories in the region. The hired weavers who worked on the almshouse factory looms could earn a "decent competancy."[62] Those who wove more than thirty yards of fine cotton cloth made well over a dollar per day in the factory. The committee raised piece rates when demand for cloth increased, as in early 1807 when a "constant call for tow cloth" compelled the managers to augment the price of weaving by one cent per yard. If the actual rate of a weaver's wage shifted with the cloth market, the amount of payment he received was determined by the quality of his work, as judged by the factory superintendent. The weavers were not paid until their products had been inspected and the proper deductions were made.[63]

A combination of the market for yarn and poor relief policy also affected the earnings and employment of the Guardians' outdoor spinners. No evidence tells us of the family economy or marital status of these women, but none received more than a bare subsistence income from her labor in the Guardians' putting-out system. They were given mostly tow or flax to spin at the rate of twenty-three to twenty-four cents per pound. Because the committee tried to provide a minimum amount of work to all who applied, few women received more than three pounds of material per week. According to the spinners accounts of 1806–7, from November through March as many as 200 spinners would produce enough flax or woolen yarn to earn typically ten or fifteen cents per day. The committee mentions in the minutes of March 6, 1811 that there always seemed to be "a good deal of demand for outdoor spinning."

The spinners' wages rose and fell with the demand for their product. In October 1807, for example, the committee had to raise the piece rate for linen and tow yarn to meet the wages offered by private manufacturers because some of their spinners were going to them for business. The invisible hand of the market turned just as easily against the interest of the spinners, as happened in June 1811. Because of an oversupply of spun tow on the market, the Guardians refused to pay the spinners for any yarn they brought to the factory. The superintendent rejected their requests for more flax throughout the summer.[64]

There is no doubt that the weavers and spinners employed by the Guardians of the Poor were in a peculiar position of dependency as wage earners. They were among the poorest of the city's laboring class. A very narrow margin existed between their earning power and entry into the almshouse as factory paupers. The experience of a few individuals portrays the difficulty of practicing a weaver's livelihood independent of the Guardians' textile factory. John Store, for example, labored as a pauper weaver for over a year and a half and then was discharged. Within two weeks he was back in the almshouse employed as an oakum picker. A few months later, he appealed to work again on the handlooms of the factory. Andrew Nelson first appeared on the factory accounts as a yarn sorter in the weaver's room in 1808. In 1811, the Guardians employed him as a hired weaver. Characterized as a workman who "does not work any scarcely," he was dismissed after one month. Nelson returned to the almshouse factory two years later, this time as one of the indoor pauper weavers. At the age of fifty, Conrad Snyder, a handloom weaver, found employment on the factory looms. After a year, however, he appeared on the almshouse admission records, listed as a former weaver and impoverished.[65]

The almshouse labor force differed from Nicholson's employees of the previous decade. Nicholson's glass workers, machine makers, and framesmiths were more highly skilled and could command substantial wages for their skills. They retained more control over the labor process and work regimen. They had the relative power to halt production and used this when conflict arose between them and Nicholson over payment of their earnings.

Though less highly skilled and with a more tenuous hold on the means of their livelihood, the hired wage earners of the almshouse also strove to keep control of the labor process. A key to this struggle again lies in their transatlantic experience. Scores of spinners and weavers on the Guardians of the Poor accounts were Irish and English immigrants who, from their homeland experience, likely associated "factory work with pauper compulsion." The first English manufactories were workhouses to employ the poor. Factories were thought of as repugnant public works and, as Arthur Redford argues, may "have been shunned originally as an insidious sort of workhouse."[66] There is evidence that Philadelphia's outdoor weavers and spinners not confined in the almshouse preferred to labor in the putting-out system and to stay out of the factory, which stigmatized them as indoor paupers and subjected them to the particular discipline of the Guardians of the Poor. Michael Roach, for example, was a stocking weaver who worked the first six months of 1811 on the almshouse factory looms. He resigned but returned a half year later requesting employment again, but this time on the terms that he be allowed to take the stocking frame "out of the Institution."[67] Factory machines often remained unoccupied. In May 1810, the committee found "several looms idle in the weaver's apartment" and ordered the superintendent "to procure good and faithful tradesmen" to weave on all the vacant looms in the factory. A week later, the factory looms were still unemployed. In the first few months of 1811, "the machinery for spinning wool and also one of those for spinning cotton" were "without employ" in the factory. During those same weeks, while the almshouse searched for skilled mule spinners to attend the factory machines, "a good deal of demand for outdoor spinning" existed and the rolls of female outdoor wheel spinners swelled.[68] Embezzlement of yarn and cloth, a not uncommon practice of the English putting-out system, was one way that these tradespeople struggled to avoid pauperism. The committee, early in January 1807, installed a pair of scales to aid in stopping the "great irregularity in the manner of delivering out the cotton and wool." But the scales did not put a stop to the embezzlement of goods. The committee discovered in 1811 that they were still suffering considerable losses because many spinners were not returning the articles that were put out to them. The supervisors attempted to take more care in determining those spinners who were "proper objects" to receive flax and wool.[69]

Those spinners and weavers who did enter the almshouse, whether as hired or pauper labor, resisted the manufacturing committee's efforts to regulate their work life and behavior. Factory supervision may have reduced the embezzlement problem of the putting-out system but factory regulation did not guarantee the committee uniform control over the production process. Factory workers defied the set of rules established by the committee to govern production. "Several persons employed in the Manufactory as well as paupers as hired people, had in some instance refused to obey the directions of the superintendent," the steward complained within a few weeks of implementing the factory regulations in early 1808.[70] After a half dozen years of operation, the manufactory committee was still dealing with the problem of hired weavers who "frequently absented themselves from work" without leave.

The years of trade restrictions and the War of 1812 halted the flow of British cloth into Philadelphia and created a high demand for American-produced textiles that lasted through 1816. In the latter part of this period of expanding demand, however, the putting-out system of the almshouse ground to a halt. As private merchant capital was redirected into textile production and drew upon the city's pool of weavers and wheel spinners, the Guardians of the Poor withdrew from yarn and cloth manufacturing. By 1811 only a handful of weavers collected piece rates from the Guardians. The almshouse continued to put its pauper female residents to work on spinning wheels, but fewer and fewer outdoor spinners received flax and wool from the factory to spin on their domestic wheels.

It is not entirely clear why the Philadelphia almshouse outwork system declined in this period. But considering the broad picture of textile production in the port city, it appears that the wartime halt in the immigrant labor supply, rising costs of raw materials, and competition from private manufacturers, some of whom adopted mechanized spinning and weaving, may have pushed the Guardians of the Poor out of the large-scale production of textiles. The almshouse factory, however, left a legacy that characterized textile manufacturing through the remainder of the decade: the employment of a large supply of skilled domestic weavers and spinners and the persistence of the traditional labor processes within a putting-out system.

Philadelphia investment capital flowed into home manufacturing, including yarn and cloth production, when the Embargo Act deadened the carrying and export business in 1808. The Mid-Atlantic port had always been the entrepôt for inland-bound manufactured goods from Europe as well as Boston, and the embargo encouraged Philadelphia's merchants to invest in production for the "daily increasing" western market.[71] The sales records of Almy and Brown, New England's oldest and largest yarn manufacturer, graph this trend. In 1806, as noted earlier, Almy and Brown sold approximately 16 percent of their coarse cotton yarn through their Philadelphia agent. By 1808, the year of the trade embargo, that figure leaped to 30 percent. The trend continued through the war years. And in 1814, when they shipped two-thirds of their yarn to Philadelphia, the New England firm testified to the potential for that city's manufacturers. "In case we send all we make, yarns and cloth, to that market," Almy and Brown wrote their Philadelphia agent, "we thinks we shall not be able to send half the amount that may be sold."[72] In that year, a Philadelphia census of the city's branches of manufactures reported that 3,071 "hands" were employed by both cotton and woolen manufactures.

Interest peaked anew in mechanization as raw material prices rose in the war years. Raw cotton that sold for seven to ten cents per pound before the war could not be found for less than forty cents per pound by December 1813. It is not surprising that manufacturers searched with new vigor for machines that would ensure the productivity of labor. In the year that war broke out, a considerable number of the 237 new mechanisms that received patents were for the production of woolen, cotton, and linen yarn and cloth. A dozen of these devices applied just to the process of spinning.[73]

Those inventors with an eye to the particularities of the Philadelphia textile trades created labor-saving machinery for the home. One Philadelphia entrepreneur introduced a "portable or family spinning machine" of ten to fifteen spindles; it could be used for either wool or cotton and took up no more space than a spinning wheel. A machine displayed at Peale's Museum in 1814 was specifically designed to increase the productivity of the female hand spinner. The inventor, John Baxter, advertised the machine for between one and three hundred dollars, depending on the number and type of spindles, clearly more than a wheel spinner could afford. Baxter promised the capitalist investor that he would clear expenses in a year because the device more than doubled the output of a wheel spinner. Only twenty-eight square inches in size, the six-spindled Baxter machine could be operated by a child as young as five, the larger machine by a girl of ten. "The child that attends them has nothing of consequence to do except turn a crank, taking off the full and putting on the empty bobbins, mending an occasional broken thread," Baxter noted for potential buyers. He perceived that a home of just four females could be turned into a profitable manufactory with a two-hundred-dollar carding engine, a fifty-dollar roving machine, and his labor-saving spinning device.[74] The introduction of such mechanisms to the Philadelphia market underscored the contrast between the urban and New England system of production, where manufacturers adopted mill machinery that used child labor drawn out of rural homes.

Simultaneously, with the movement to increase the productivity of the urban female spinning force, American textile manufacturers turned their effort to replacing the skilled handloom weaver with unskilled female labor. In the first year of the War of 1812, eight or ten patents were issued for different looms designed to mechanize various hand processes of the skilled weaver. Inventors of weaving machinery plied the Philadelphia market. One wartime letter to a Philadelphia resident suggests that a number of labor-saving looms were being shown to the city's merchant-manufacturers. A New Jersey promoter urged his Philadelphia friend not to purchase "any of these spurious things that are carried about" until he viewed his Connecticut loom. The weaving process of this machine was mechanized to the extent that "by merely moving the batten or lathe, every movement necessary to any loom is effected." Supposedly a child of ten or twelve could be taught in just a few days to operate the loom. "The quantum done is about double that of the common loom, with less than half the labor," the advertiser promised.[75]

By the end of the War of 1812, however, dealers of the labor-saving looms could see that Philadelphia was not the place to make a sale. The story of the successful introduction of the powerloom in 1814 and its widespread adoption belonged to New England. Fed by a new postwar influx of immigrant tradesmen, the sheer number of Philadelphia handloom weavers and the low price of their labor stalled the general adoption of the powerloom in the port city until after the postwar depression.

On the other side of the Atlantic, early nineteenth-century handloom weaving offered a marginal livelihood and status. We have seen that Ulster linen weavers and other coarse cloth producers had flowed into Mid-Atlantic American ports prior to the turn of the century. In 1810, however, trade societies for the first time encouraged weavers to emigrate. Philadelphia again received many of those who left when the nonimportation policy of 1811 compelled American vessels, forbidden to carry cargo, to transport emigrants between Atlantic ports. After peace in 1815, British soldiers and sailors "exchanged the musket for the shuttle" and "joined the already overcrowded ranks of weavers." As wheat prices and unemployment rose and wages dropped, unprecedented numbers of displaced Irish and English tradesmen left their homelands for America.[76]

Though strapped by its own postwar economic slump, Philadelphia became the destination of thousands of Britain's "best hands in arts and manufacturers." A British newspaper lamented in 1816 that "the continued and increasing emigration from this country to America becomes every day more alarming."[77] Dozens of vessels sailed weekly from Londonderry, Belfast, Dublin, St. Andrews, Newry, and Liverpool. An average of 200 to 290 persons from Irish and English ports disembarked every week in Philadelphia and New York in the summer months of 1816, 1817, and 1818 when the standard fare was around ten guineas. Within one week of August 1816, two Londonderry ships arrived in Philadelphia carrying a total of 199 passengers.[78]

Like their predecessors of two decades before, many of the postwar immigrants settled in the port city of their arrival. In the summer of 1816, the Baltimore-based *Niles' Weekly Register* bemoaned the nature of the new urban settlers: "One Swiss is worth a hundred of the Cockney tape-sellers with which our cities have teemed." Philadelphia's Irish population was large enough by then to support three Irish benevolent societies.[79] We do not know the exact number of male weavers or female and male spinners who immigrated to Philadelphia after the war. It is clear, however, that the reorganization of the Irish linen industry alone was an unremitting cause of the postwar emigration of the farmer-weavers of rural Ulster and the cloth producers of the cities. William F. Adams, a historian of Ireland's emigrants, argues that "the greatest emigrant counties in Ireland were those largely inhabited by the small independent weaver."[80]

Philadelphia textile capitalists must have observed with particular interest and perplexity the swelling pool of skilled handloom weavers. For at this very time trade protectionists were expounding with new enthusiasm that women, not able-bodied men, could and should attend the machines of domestic manufacture. The manufacturers of textiles in New England reported to a congressional committee in 1816 that they employed women and children in all their factory departments except the machine shop.[81] Boston and Providence, however, had received but few of the labor-laden vessels from Irish and English ports. Only nineteen ships from Irish emigrant ports had landed in Boston between 1815 and 1819, compared with sixty-two in Philadelphia. Indeed, the development of the industry in Philadelphia, which became the home of so many British weavers, mirrored the English experience, where "the very cheapness and superfluity of handloom labour retarded mechanical invention and application of capital in weaving." While New England manufacturers, beginning in 1814, eagerly adopted the powerloom, which rural women and girls operated, their Philadelphia counterparts began to gather the less expensive handlooms in centralized factories to capitalize on the low cost of their skilled urban work force.[82]

Although powerloom weaving in Massachusetts was not replicated in Philadelphia, the New England industry pushed independent Philadelphia handloom weavers to the specialty market and paved the way for Philadelphia capitalists to invest in handlooms and buy the labor to operate them. The technology of the early New England powerlooms limited that region to the production of coarse, plain cloth. Because of the mechanical regularity of the beating-up process, these New England looms produced a better quality, as well as higher quantity, of this coarse cloth than handlooms could. The mass production of high-quality sheeting in New England factories drove Philadelphia weavers out of the plain cotton line and into the production of higher-priced colored materials. The demand through the end of the war for the more expensive plaids, stripes, ginghams, and shirtings that the powerloom could not weave put an increasing number of mule spinners, wheel spinners, and weavers to work in Philadelphia's cotton textile industry. One survey of manufacturers in Philadelphia reported that the number of hands employed in cotton manufacturing had increased 32 percent just between 1814 and 1816.[83]

The manufacturing census schedules of 1820 provide a glimpse of the work places of these postwar Philadelphia weavers and spinners. In this period, textile production was in a transitional stage. Yarn and cloth production was moving out of the putting-out system and into centralized manufactories but not out of the hands of skilled weavers and mule spinners. While production became centralized, few of the owners of the textile firms had invested in powerlooms and water-powered spinning frames. Of the thirty-five textile manufactories that returned schedules in Philadelphia County, five were cotton-spinning mills, twelve were combined spinning and weaving factories, and the majority (eighteen) were exclusively wool- and cotton-weaving establishments. The thirty cloth manufacturers gathered from half a dozen to forty handlooms and weavers in their establishments. Only two reported operating powerlooms. The manufacturers of yarn used the traditional mules and jennies. Only Henry Whitaker on Frankford Creek enumerated a water-powered spinning throstle among his machinery.[84]

An observer would have counted few women entering one of these textile manufactories that dotted the city's southern and northwestern perimeters. Mostly children attended the carding machinery and assisted the male mule and jenny spinners and weavers in the combined spinning and weaving establishments. Sam Haydock's schedule reveals that his Adelphia Cotton Mills in Blockley was an exception, employing thirty women and girls, an unusually large number. He had invested $40,000 in mill seats and machinery, which included three large powerlooms.

The size and structure of the male handloom weavers' workplace varied between the combined weaving and spinning factories and smaller weaving shops, mostly located in the southern districts of the city. In Moyamensing, for example, master manufacturers owned less than seven handlooms and did no spinning in their establishments. Whether they operated four or forty looms, these manufacturers of cloth identified themselves as weavers by trade and training. They needed relatively little capital to invest in a handloom, which cost between $60 and $150. The majority of their outlay went into paying the wages of fellow weavers.[85]

The small weaving shops outnumbered the joint spinning and weaving factories. The greatest number of weavers, however, labored alongside male mule and jenny spinners in the large factories that were concentrated in the northwest section of the city. Twenty-two to 30 handlooms could be found in five of these textile mills. The Globe Mill in Kensington operated 140 handlooms along with 3,200 spindles. In the heart of the weaving district, this factory was by far the largest single employer of factory textile labor. In addition to the factory loom tenders, 150 independent weavers in the neighborhood bought and wove the surplus yarn that the Globe Mill spinners produced. When demand for goods was high, the yarn produced at the factory provided work for 530 Kensington spinners and weavers. Much more capital was invested in the mill and machinery of even the smallest of these textile manufactories that combined weaving with spinning on mules and jennies than in the exclusively handloom manufactories. Indeed, none of the owners of the mills in the northwest section of the city were weavers by trade. It was merchant capital, as much as $130,000, that brought male weavers and spinners and their looms, mules, and jennies into the factory system of production.[86]

The postwar economic depression, which temporarily crippled the port city's textile industry, doomed the small handicraft producer and small shop owner. The depression had a particularly debilitating effect on spinners and weavers of fine cotton yarn and cloth. The New England factories were not adversely affected by the peacetime flood of fine English cloth because they manufactured the cheap coarse cloth. They alone received protection from the 1816 tariff that excluded the coarse Indian cottons—the only type that competed with the powerloom sheeting. Philadelphia textile producers of ginghams, shirtings, and calicos could secure no such restrictions on the "unusual quantities" of fine foreign goods with which they competed.[87]

Between 1816 and 1820, piece rates for handloom weaving plummeted by one-third to one-half as large quanities of high-quality English cloth sold on the city's auction blocks. A committee investigation of the situation of the Philadelphia manufacturers reported that the number of hands employed in cotton manufactures had dropped 93 percent between 1816 and 1819.[88] In 1820, few of the owners of manufactories of either yarn or cloth did not complain of the "heavy importation from Europe" or "influx of British goods" that had reduced prices and deadened business. Owners of cotton-spinning mills simultaneously faced a 20 percent advance in the price of raw cotton. Almost all manufacturers of yarn had thus reduced their spinning force and machinery in operation by one-third to one-half. "These circumstances prevent any chances of profit," the merchant-turned-cotton-yarn-manufacturer David Lewis complained to the census taker. Those who manufactured cloth had to sell their goods at auction. "All I make is by boarding my workmen," complained Joseph Crosby, the owner of twelve looms.

While large-scale manufacturers of yarn and cotton cloth were forced to curtail production, none felt they would be driven altogether out of business. By contrast, the bitter testimony in the 1820 census of the weavers who owned the small handloom shops bore out the ruinous impact of the depression on those with little capital. Half of John Hinshilwood's fourteen handlooms lay idle in 1820, and he had reduced his weavers' wages from between seven and eight dollars a week to four dollars. If the depressed market conditions continued, he wrote, "I'll be obliged to drop it altogether.... We have no other market but auction houses." Virtually all of the weavers who owned half a dozen or so looms recorded on the census questionnaire that they were going out of business. "I have little to say or wish to say little," James Maxwell replied "as we unpro[tected] have to contend with Foreigners Protected." Tradesmen James Stranaghan and Andrew Kilpatrick both believed that it was only a matter of weeks before they would have to close their weaving shops. And Edward Ervin and John Murphy, weavers in Moyamensing, voiced the same complaint that "the weaving business is at Present that a man May Live but very Poorly."

The depressed cloth prices in these years no doubt caused more and more looms to fall into the hands of capitalist manufacturers. At the end of the cycle of economic stagnation and recovery, a number of these capitalists began to rent waterpower and invest fixed capital in spinning frames and powerlooms and in the mill seats on the newly constructed canal on the banks of the Schuylkill River in Roxborough Township.

In exploring the transition of labor and capital to the mechanized factory, it is vital to recall that for at least three decades the relations of the textile laborers and capitalists had developed within an outwork and manufactory structure. The relatively late adoption of mechanized factory production, in contrast to New England, came *after* the traditional processes of weaving and spinning had taken root, merchant weavers and yarn manufactures had accumulated capital, and the sources of raw materials and markets had been well developed. Central to this particular course of textile production in the Mid-Atlantic seaport was the labor supply, a constantly replenished pool of skilled immigrant weavers and spinners. Their poverty underlay the birth of the textile manufactory in the workhouse, and their skill made Philadelphia a center of fine yarn and cloth production. For thirty years, the immigrant handloom weaver and mule spinner, in particular, had delayed the coming of mechanization. Thereafter they would shape the social conflict that accompanied it.

III

Mechanization and Mill Production in the Urban Seaport, 1820–1837

In the 1820s, Philadelphia's textile industry, which had long resisted mechanization, made the move into the water-powered mill and the era of industrial factory production. This process, unfolding rapidly, took place along a recently completed section of the Schuylkill Navigation Company canal in the township of Roxborough. By 1828, the newly named mill town of Manayunk was being likened to Lowell and Manchester. The transition to the mechanized factory marked a decisive change in the relationship of the urban textile worker and manufacturer. Indeed, the social relations of the mill began to forge mill workers and mill owners into Philadelphia's early industrial working and capitalist classes.

The large-scale mechanized production of cotton textiles developed, not in the traditional centers of cloth and yarn production in the port city—Kensington, Moyamensing, and the Northern Liberties—but in Roxborough's village of Manayunk. The growth of mills and the adoption of labor-saving machinery in this factory town, located six miles from the city center, turned Philadelphia into the leading textile-producing region in the country in the antebellum period.[1] The regimen of factory production that took hold in Manayunk contrasted sharply with Nicholson's fitful system and with the unsupervised outwork system of mule spinners and handloom weavers. We can sense the pace and magnitude of change that accompanied mechanization through a glimpse at the number and size of the mills that arose on the banks of the Schuylkill during the 1820s and early 1830s.

Manayunk became the home of Philadelphia's early factory system in part because the Schuylkill Navigation Company realized that their canal was suitable for the development of mill seats. The company correctly anticipated that it could gain profitable yields from the sale of waterpower for manufacture and the leasing of its lands for factories and dwellings. In 1819, they finished construction of the two-

Mechanization and Mill Production, 1820–1837

mile-long Flat Rock Canal through Manayunk and a dam that backed water up the Schuylkill for four miles, producing a fall of twenty-six feet from canal to river. The canal company began selling this water power for $3.00 per inch per year, and by the end of the decade, it collected $12,000 annually from the Manayunk mill operators.[2]

The building of mills and installation of labor-saving machinery followed quickly behind waterpower sales. "Mill is huddled on mill," a correspondent to the *Democratic Press* wrote of Manayunk in March of 1823. By the summer of 1824, there were seven mills and factories on the canal, the largest of which was a cotton mill containing 1,500 spindles and 60 powerlooms that was expected to employ 200 hands.[3]

The heyday of mill building came in the late 1820s when the foreign markets for coarse cloth grew tremendously. "The importance of labour-saving machinery is gaining ground daily," reported the *Mechanics' Free Press* in August 1828. By that summer, the borough was the home of five large cotton mills and two woolen factories, with more factories under construction.[4] In terms of the number of spindles, powerlooms, and hands employed, Manayunk was now larger than Pawtucket, where Samuel Slater had laid the foundations for New England industrialization forty years before. Five hundred and twenty-five hands attended 14,154 mule and throstle spindles and hundreds of powerlooms. Such statistics gained Manayunk the title "the Lowell of Pennsylvania." Yet a more fitting comparison, as we shall see, was made by Manayunk's Fourth of July orator in 1828. "We have received the appropriate and highly complimentary appellation," he boasted, "of the Manchester of America."[5]

Mill building on the canal in Manayunk proceeded steadily into the next decade. "In the last fifteen years," a Manayunk booster announced in 1834, "mills have been erected; a large amount of capital invested and a large amount of labour-saving machinery introduced." The 1840 census of manufactures documented the concentration of Philadelphia's coarse cotton textile industry in Roxborough Township. Manayunk's eight mills, which represented only 18 percent of those in the county, operated 44 percent of the spindles and contained 27 percent of the invested capital. Over a quarter of all the county's textile operatives labored in the factories of this industrial town.[6]

Why did the industrial mill gain such a firm foothold during the decade preceding the Panic of 1837? Certain historical factors converged to induce Philadelphia textile capitalists to adopt, in rapid fashion, the New England system of water-powered cotton textile production in the 1820s. First, the expansion of the foreign demand for coarse cotton goods coupled with the revival of the domestic southern and western market promised for the first time to sustain

A serene view of the mills and town of Manayunk on the Schuylkill River, J. T. Bowen, Philadelphia, 1840 (Historical Society of Pennsylvania).

the large-scale production of powerloom cloth in both Philadelphia and New England. Second, English and Irish emigrants from Britain's stagnating textile districts flowed into the Mid-Atlantic seaport in unprecedented numbers in the late 1820s, broadening the supply of unskilled female and child laborers. Third, and perhaps most important, technological improvements in the powerloom and the spinning throstle and their diffusion made it imperative that Philadelphia textile capitalists invest in the labor-saving devices in order to compete with New England and survive in the manufacture of cotton textiles. The decision to mechanize seems to have been affected also by the growing militancy of skilled mule spinners and handloom weavers. The influence of these factors on Philadelphia's early industrial history deserves elaboration.

In the early 1820s, the domestic demand for the tariff-protected New England power-woven cloth, "increasing each season," certainly caught the notice of Philadelphia entrepreneurs. The port city still predominated as the entrepôt of western and southern-bound cotton goods. The Slater Company in Providence consigned over 85 percent of its coarse cloth to their Philadelphia agent in 1821. Agents for the Poignand and Plant Company of Massachusetts wrote that everyone in the Mid-Atlantic city was eager to place large orders, for the coarse cotton goods were now in demand all over the "Carolina and western countries."[7] Dozens of Philadelphia cotton manufacturers entered the large-scale production of coarse goods during this early postdepression boom. In 1824, however, the domestic market collapsed as beleaguered British manufacturers dumped low-priced goods on the overstocked American market. The *Niles' Register* portrayed the Philadelphia textile manufacturers as victims of the competitive market system: "Capitalists were induced to invest their money so plentifully and spindles and looms multiplied so rapidly that the consumption could not keep pace with them."[8] Since domestic competition was substantial, Philadelphia cotton cloth manufacturers eagerly turned to foreign outlets at mid-decade.

By the late 1820s, American cotton textile capitalists rivaled the British with their distinctive brand of "coarse and stout manufactures," commonly known as "domestics" or "domestic goods." Mexico, the West Indies, the new republics of South America, as well as the East Indies, imported bales upon bales of Philadelphia-produced domestics—coarse sheetings and shirtings—in the mid-1820s. By 1827, Philadelphia manufacturers had expanded their international markets, sending large quantities of cotton cloth—"the cheapest and best in the world"—to Canton and the Mediterranean.[9] The overseas demand for these coarse goods sustained Philadelphia and New England mill owners through the early 1830s. James Kempton, a textile manufacturer from Philadelphia, testified before the British Parliament in 1833 that his country's production of coarse cotton cloth had increased 600 percent in the previous sixteen years, a growth rate almost quadruple that of Great Britain.[10]

The healthy worldwide demand for the product of the American cotton mills allowed some Manayunk mill owners to challenge the corporate mills of Lowell in size and output. J. J. Borie was one such large-scale cotton manufacturer who entered the ranks of the early industrial capitalists with no prior manufacturing experience. A French refugee from Saint Domingue, he arrived in Philadelphia in 1805. With a fellow emigré, Peter Laguerinne, he established and developed "one of the most important commercial houses in the U.S.," importing from Europe, China, and India, and exporting dry goods to China, South America, and Mexico.[11] Having garnered a great deal of wealth in the marketing of coarse goods, Borie decided to invest in the production side of the business. In partnership with Laguerinne and Jerome Keating, the nephew of another French refugee, Borie built a $120,000 cotton mill on the Manayunk canal in 1828, the whole cost of which, Borie would later boast, was paid in cash. He assured his insurance company that "no money or any other thing has been spared in the completion of this manufactory."[12] Comparable in size to one of the Lowell mills, the four-story, 200-foot building contained 4,500 spindles and 120 powerlooms. Two hundred and fifteen mill operatives produced 20,000 yards of coarse cloth per week for the export market. Borie and Laguerinne's merchant house sent brown sheeting, a heavy cheap cloth, to agents in Vera Cruz, Tampico, Buenos Aires, and to Havana for the Spanish slave trade.[13]

Borie's report to Henry Clay in 1832 that his was the "largest cotton mill in Pennsylvania" suggests the relative success of this merchant-turned-manufacturer in the large-scale production of cotton cloth. A neighboring Manayunk mill owner, Joseph Ripka, however, could rival his claim. Rising through the ranks of the master manufacturer, Ripka represented a different and more dramatic transition to industrialist capitalist. A weaver by training, he abandoned a small cloth industry in Lyons in 1815 and immigrated to Philadelphia where he set up a modest handloom shop in Kensington. His weavers produced a high-quality cassimere, and as the domestic demand for quality goods expanded after 1821, he increased the number of looms in operation. In 1828 Ripka, as Borie had done, invested his accumulated capital in a Manayunk mill and began the powerloom production of cottonades, a heavy cotton cloth used for cheap men's clothing and in great demand by southern slaveowners. By 1832, Ripka's mill was about 50 percent larger than Borie's. His Silesia Manufactory, named after his birthplace, employed 300 hands on 7,176 spindles and 224 powerlooms. Before the end of the decade, Ripka owned four mills in Manayunk and operated one of the largest textile businesses in the country.[14]

The success of the early Philadelphia capitalists like Borie and Ripka depended on both competitive and cooperative tactics. Borie demanded that his agents keep secret the information on the type of yarn or cloth they ordered or received from his Manayunk mill. "Any competition would destroy our prospects," he wrote to his Havana representative. On the other hand, Ripka and other manufacturers of cotton goods in Philadelphia established two merchant houses to sell their products collectively.[15] Whatever strategies or networks they used in the marketplace, Borie and Ripka could not have successfully manufactured cotton textiles without access to cheap labor and the constantly improving technology for yarn and cloth production.

MECHANIZATION AND MILL PRODUCTION, 1820–1837

The cotton mills of Joseph Ripka seemed to dominate the landscape of Manayunk in the late antebellum period, W. H. Rease, printed by Wagner & McGuigan, Philadelphia, 1856 (The Hagley Museum and Library, Wilmington, Delaware).

While the foreign cloth market encouraged large-scale production of coarse goods in the 1820s, a growing population of urban poor women and children provided a reserve pool of cheap unskilled labor that encouraged textile capitalists like Ripka and Borie to mechanize. As Philadelphia's population grew in the first half of the nineteenth century, so too did the proportion of laboring poor. The depression following the Napoleonic Wars pushed the city to unheard of levels of poverty and unemployment. The number of persons in the port city applying for public relief more than doubled between 1810 and 1820 and remained high through the next decade. One of the widely varying estimates of those without jobs in the city of 100,000 was 20 percent in the early 1820s.[16]

The pattern of Philadelphia immigration described in chapter 2 continued, and refugees from the distressed textile districts of Ireland and Britain expanded the city's poor population. A brief lull in the transatlantic flow to America in the early 1820s ended when almost 8,000 Scots, English, Welsh, and Irish arrived in northeastern ports in

1825. The following year, a Liverpool official reported that the steerage passengers leaving his port were chiefly going to New York and Philadelphia. In 1828, on the eve of a recession that struck hard at the American cities, 17,840 Irish and British disembarked in the coastal ports. During one week of the summer of that year, Philadelphia received nearly 400 immigrants who had set sail from Liverpool and Londonderry. They came from Ireland, Yorkshire, Lancashire, and Scotland, where the transition from handloom to powerloom weaving and the check in the cotton trade had created "states of distress bordering upon actual famine." The peak in the flow of the job-seeking immigrants from the period between 1790 and 1840 occurred in the few years preceding the Panic of 1837. In 1834, just under 35,000 British and Irish immigrants entered the American seaports.[17]

More of these immigrants disembarked and remained in Philadelphia than jobs existed for them. Their plight, in fact, became the subject of one of Mathew Carey's impassioned essays. Portraying the distress of recent arrivals in 1826, a recession year, Carey wrote of the "great numbers crowding into" the city's manufactories, of weavers "dispers[ing] themselves through the country in search of employment," and of numbers of desperate immigrants getting back on ships to return home.[18] Unable to find "laboring work," hundreds of immigrants placed an unprecedented burden on the city's poor relief institutions. So great was the number of jobless immigrants in the seaport that the Guardians of the Poor proposed a head tax on all incoming passengers to help offset the additional cost of poor relief. Reports from the Hibernian Society confirmed that immigrants to the Mid-Atlantic port were augmenting a reserve urban labor pool. Established in 1791 to assist the Irish population, this society for relief of emigrants from Ireland filed dismal accounts in the mid-1830s of the difficulty of finding employment for the scores of destitute Irish in the city. While the turn-of-the-century textile immigrants forestalled mechanization, the post-1820 wave facilitated the development of the mechanized factory because a larger proportion of the Atlantic migrants were unskilled women and children.[19]

Among the urban poor of the 1820s and 1830s, able-bodied women figured importantly. According to one study, nine out of ten persons receiving outdoor cash aid in the 1820s were females, and a third of this number were mothers, most of whom were widows with young children to support. The vast need for jobs among Philadelphia women is further suggested by the establishment in 1824 of the House of Industry of the Provident Society. Reaching back to a traditional form of urban textile production, this institution provided outwork for hundreds of "industrious poor women" in the city. In 1829 the federal government took advantage of the port's reserve of female laborers, employing 400 of them to sew shirts and pantaloons for the army at a weekly rate of $1.31.[20]

By 1830 the "oppressive superabundance of female labour" in Philadelphia had become a cause célèbre of Mathew Carey. Through the 1830s, Carey "repeatedly pressed on the public attention" the miserable conditions of the seaport's laboring women. Echoing the moral rhetoric of the postrevolutionary proponents of manufacture, Carey and fellow crusaders called for putting women to work in manufacturing production. Appalled at the "poverty and destitution" of urban females, a Baltimore correspondent of Carey's proposed that men should abandon all the "spheres of employment [within] the limits of [a woman's] strength." Carey himself saw the cotton mills of Philadelphia as a remedy for the moral and economic problem of female poverty in the city. Given "the want of females [employed] in factories as spoolers, spinners, and weavers," the public-spirited elite should send the "superabundant poor families of our cities to these places," he argued.[21]

Carey's *Appeal to the Wealthy of the Land* not only underlined the need of Philadelphia's laboring women for employment but also pointed out the demand of the cotton textile manufacturer for the labor of such women. Another keen observer pinpointed the cost-conscious manufacturer's logic of using Philadelphia's particular labor reserve. Such women "could find no other employment and therefore would have no choice but to submit to [a] reduction of wages which although it made their situation worse, would keep them better than they would be without any employment at all."[22] The Philadelphia textile capitalist understood that the city's poor women and children provided the cheap labor supply that was essential for investing capital into fixed machinery and mills.

The poor urban family became an important unit of labor for the Philadelphia and Manayunk textile mills in the 1820s. Child labor and the family employment system were not, however, an altogether new form that came into being in the industrial factory. Philadelphia yarn manufacturers had hired mule spinners as subcontractors in the decades when Philadelphia became a center of fine yarn production following the turn of the century. The skilled mule spinner received a family wage from which he would pay his own piecers and assistants. Half of the twenty-two men and women employed in Henry Whitaker's mill in Oxford in 1820 and 1821, for example, had children who assisted them as piecers or bobbin winders. Five fathers and five mothers each collected the wages of one to four children.[23] The postdepression surge in coarse cotton textile production, however, created an unprecedented and, at times, seemingly insatiable demand for children and young women as unskilled machine attendants.

Evidence of the transformation of Philadelphia's poor children into a segment of the urban industrial laboring class appeared soon after the revival of the economy in 1822. In that year, and again in 1823, the controller of the Philadelphia schools complained of the diminishing number of pupils because of the "employment which is given them in the numerous manufacturing establishments within the district."[24] In 1828 half of the 1,099 inhabitants of the industrial textile town of Manayunk were children under fifteen. By the end of the decade, mill owners in Manayunk and elsewhere met their growing need for children and women laborers by actively recruiting urban families for their factories. For example, on August 7, 1830, the Good Intent Factory across the Delaware River in New Jersey advertised in Philadelphia's *Mechanics' Free Press* for eight or ten female powerloom weavers. The advertisement specified a preference for a family that could furnish four or five hands.[25]

In the early 1830s, when the demand for labor was great in the textile districts, mill owners gave preference both for jobs and rental dwellings to the parents who could supply the greatest number of laborers. In April of 1834, when J. J. Borie placed an advertisement in Philadelphia's *U.S. Gazette* for "ten or fifteen families" to labor in his Manayunk mill, he added that "the larger the families the sooner they will be employed." Following the 1837 depression, Joseph Ripka had to recruit actively in the port city for families who could place members in the powerloom or card rooms. Some Manayunk mill owners apparently filled their labor needs by less scrupulous methods. The principal of the public school in the textile town condemned the mill owners before a state investigative committee for frequently sending mill bosses or spinners to snatch young scholars "from their seats."[26]

The accounts of the demand for large numbers of textile machine attendants should not disguise the serious need of the mill laborers for such employment. The poverty of the mill workers and their families, which also characterized the early Rhode Island textile industry, was indeed a mark of the urban textile labor force. In the *Germantown Telegraph* of January 1, 1834, a Manayunk resident reported that the wages and employment opportunities for children had brought 3,000 persons to the textile district. "By far the greater proportion," he pointed out, "were large families in very moderate circumstances." Three years later, the state's first investigation into the factory system in Philadelphia and Pittsburgh exposed the grimness of the "moderate circumstances" of the mill laborers. Witness after witness described the desperate need of the mill families for employment. They all agreed that the urban poor largely provided the labor for the mills. When downturns in the market economy lessened the mill owners' demand for a labor force, poor parents would take their children from mill to mill begging for employment. John Thornily, who had worked as a foreman and machinist in Pennsylvania mills for seventeen years, made the statement, echoed by many witnesses, that widows and widowers commonly reared their families in the textile mills. If the state legally prohibited children from working, the witnesses concurred, their parents probably could not sustain themselves and their families.[27]

Dire poverty and the family employment system distinguished the experience of the Philadelphia textile mill worker from that of the "very lady-like females" who filled the mills of Lowell.[28] The background of the Manayunk labor force differed significantly in another way from that of their southern New England counterparts: most were Irish and English immigrants. Whereas foreign-born workers composed a very small proportion of New England mills before 1840, mills in the Philadelphia region were filled with immigrants who had been disembarking in significant numbers since 1825.[29] In 1833 James Kempton, a Manayunk mill owner, told the British Parliament that a "great many persons from England were in the cotton mills of Pennsylvania and New Jersey." And a few years later, Charles Hagner, a fellow Manayunk manufacturer, blamed the "new population" in the textile town for introducing the custom of child labor to Philadelphia's textile factories.[30]

It was not the expanding labor reserve or markets alone that encouraged Philadelphia capitalists to invest in labor-saving looms and throstles, but also, in the words of one Manayunk manufacturer, "the almost daily improvement in machinery."[31] Innovations in the spinning throstle and powerloom in the late 1820s gave the new technology three compelling advantages over the traditional processes of mule spinning and handloom weaving: lower labor costs, greater productivity, and increased control over production for the capitalist.

The third point should be underscored. Philadelphia textile manufacturers desired to eliminate their dependence upon male spinners and weavers because they were more independent as well as more expensive than factory operatives. E. P. Thompson's statement that "manufacturers in the first half of the nineteenth century pressed forward each innovation which enabled them to dispense with adult male craftsmen" applied to Philadelphia as well as to English cotton manufacturers. When James Kempton of Philadelphia testified before Parliament in 1833 that "we have as few men as possible in our manufactories," he was describing a strategy both to utilize cheap female and child laborers and to circumvent the militancy of the urban tradesman.[32]

In the early 1820s, the numerous Philadelphia mule spinners spun all the cotton yarn that was used for the filling or weft of the weaver's looms, just as they had in the previous decades. The mule spinner, as noted earlier, was highly skilled—the intermittent action of the mule demanded dexterity and strength—and he was the highest-paid textile producer in the Philadelphia industry. While the spinning throstle came into wide use in the mid-1820s because of the expanding market for cloth, the mule spinner was not threatened because the process of water-powered spinning was not yet adaptable to fine yarn spinning.[33]

Within a few years, however, American textile capitalists began to seek technological improvements that would eliminate the costly and independent mule spinner. In 1827 the Nashua Company of New Hampshire designed a mule spinning frame to produce the better and softer cloth of the mules without the skilled spinner. "In this country," the Nashua Company directors reported, "the expense of employing mule spinners and the extreme difficulty of obtaining good and faithful ones have caused the mule to fall into general disuse in the large factories."[34] The company was pleased to announce that their improvement would "wholly dispense" with the male mule spinner. Girls and boys would be able to tend the new mule frame. Despite such optimism, the early mule frames at first proved unsuccessful.

In 1828 a New England inventor, Charles Danforth, devised a throstle that could compete with the mule in the production of fine cotton yarn. Danforth made a simple improvement in which the bobbin, instead of the spindle, rotated on the frame. The bobbin dragged the thread, the reverse of the old throstle, allowing a uniform tension and giving the yarn greater evenness. The device not only produced "the finest yarns" but exceeded the output of the mule by 40 percent. By the mid-1830s, a female spinner in Philadelphia could operate a throstle that turned out one-and-a-half times the amount of fine yarn as the standard mule. She labored for $2.25 per week, compared to the male mule spinner who earned between $6.50 and $7.50 per week.[35]

The advantages of productivity and control made the Danforth frame extremely attractive to manufacturers. This was evidenced in the conflict that surrounded the efforts of J. J. Borie to replace his mule spinners with the new female-operated machinery. In January 1831, Borie wanted to increase the output of his yarn department and wrote to machine shops in Paterson, New Jersey, in search of someone who could build Danforth's frame. The Manayunk mill owner felt confident that the improved throstle machine "was destined to supercede mules for such numbers as they can spin."[36] By March, Borie had found a machine shop that could make the Danforth frame and ordered three at a cost of $288 each. He placed the order, however, on the condition that the contract be kept secret. Borie feared that his mule spinners would discover that "we have contemplation of getting rid of our mules." He had good reason to anticipate that his skilled hands might "take it into their hands to refuse working for us or otherwise molest us." Two years before, Borie's mule spinners had struck over a wage reduction and then threatened the lives of the strikebreakers in order to protect their jobs. Moving quickly, Borie succeeded in installing nine frames, apparently without opposition from disgruntled employees.[37]

The skilled mule spinner did survive in the Philadelphia industry, spinning the highest grades of cotton yarn until the self-acting mule was introduced in the 1840s. Nevertheless, by the late 1820s, his livelihood was being threatened by the textile capitalists' efforts to mechanize production. Manufacturers like Borie competed effectively in the coarse goods market because they invested in the wave of innovations that reduced their dependence on the "highly skilled, highly paid, and often independent-minded adult male spinners."[38]

The factors that shaped the course of mechanization of yarn production also influenced the introduction of the powerloom for weaving. The influx of British and Irish immigrants made handloom weaving one of Philadelphia's largest trades. Despite the abundant supply of handloom weavers, Philadelphia's textile manufacturers turned to powerlooms because, like the spinning throstles, they reduced costs, increased productivity, and wrested control of production from the skilled male laborers who were "their own masters." The wages of a female powerloom attendant were half those of a Philadelphia male handloom weaver, waterpower was plentiful and very cheap, and the powerloom was clearly a more productive machine than the handloom by the late 1820s.[39] A handloom weaver at this time could produce eighteen yards of gingham or check per day and commanded between $4.50 and $5.00 per week. In contrast, a Manayunk mill girl in the late 1820s, laboring for $2.50 per week, could tend two or three powerlooms that each produced twenty-eight yards of cloth per day. By the early 1830s, a female machine attendant could manage as many as six powerlooms, which could each turn off forty yards of cloth per day. By this time, the handloom weaver and the whole production and marketing system associated with that weaver could no longer compete in the coarse cloth market. By 1834, the combined use of the powerloom and spinning throstle in the water-powered mills of Manayunk had reduced the cost of spinning and weaving sheetings to five cents per yard, which was three cents cheaper than the cost for weaving alone.[40]

In the production of higher-quality cloth, technology steadily reduced costs and undercut handloom weavers' prices. In 1830 Alfred Jenks, a protégé of Samuel Slater, introduced a powerloom for weaving checks in Kempton's mill in Manayunk. Within a few years, the price of Kempton's powerloom checks matched handloom prices and approached the cost of plain brown sheeting. The development of calico printing, carried on in a handful of Philadelphia establishments, also brought cheap powerloom products into direct competition with the colored goods of the handloom weavers.[41]

Tension underlay the introduction of the powerloom in the same way that it shaped the course of mechanization of yarn production. The handloom weaving economy provided fertile ground for discontent. The decade was a volatile time for handloom weavers, even aside from the deleterious effect of the introduction of the powerloom. Recurring recessions in 1826 and again in 1829 threw thousands of weavers out of work in Philadelphia.[42] Like skilled mule spinning, handloom weaving continued to employ hundreds of urban dwellers throughout the first half of the century. In 1832, Henry Carey correctly predicted that the "great influx of migrants from Ireland" would keep the population of weavers "high for a long time" in America's coastal cities.[43] As the technological innovations of the powerloom steadily reduced their command of the market and their earning power, Philadelphia weavers, as we shall see, took collective direct action against mill and powerloom owners of Manayunk.

The change from outwork or manufactory production to mechanized mill production sharpened the opposing interests of urban labor and capital. The new factory system that emerged in Manayunk in the late 1820s altered significantly the position and relative power of employer and employee that had characterized John Nicholson's manufactories in the 1790s or the putting-out network of the Guardians of the Poor after the turn of the century. By examining closely the social relations of production and the conditions of labor in the mechanized mill, we can understand why Manayunk in 1828, as opposed to Lowell or Pawtucket, better deserved the title of "the Manchester of America."

The relationship between labor and capital changed in the industrial system because mechanization radically altered the world of production. Mechanization facilitated the power of the employer over production, imposing on all workers, two-thirds to three-quarters of whom were women and children, "the rhythms and movement of the machine process." In his description of the synchronized motion of every shaft and belt in the mill, Anthony Wallace has cleverly portrayed the mill as a gigantic mechanical organism.[44] The job of engineer was created out of the need to keep the machines running a set number of hours and minutes per day. "All the operatives [including the mill spinner] must keep pace with the speed of the machinery," the senate investigative committee of 1837 noted in their summary report on Pennsylvania's cotton factories. The industrial factory laborer was "regulated by the operations of the machinery." Mechanization had turned the traditional relationship of the worker and the tools of production on its head.[45]

It was the conditions of factory production that mill owners created around their water-powered machines, not the transforming effect of mechanization on the process of human labor, that distinguished the Manayunk mills from southern New England mills. In a report dated November 7, 1828, the *Mechanics' Free Press* called the mills of Manayunk "a deplorable example of the oppressive and degrading effects of machinery on the productive classes." The newspaper pointed to a system of child labor, debilitating and dangerous working conditions, and long and irregular hours of work—the "daybreak to dark night system."[46]

Child labor was one reason, though one ignored by the zealous community boosters of 1824, that Manayunk deserved the title of "the Manchester of America." "Look at Manayunk," cried the *Mechanics' Free Press* exposé, "the heart sickens to behold the remorseless system of infant labour obtaining foothold upon our soil!" Middle-class reformers were particularly concerned with the long-range effects of the factory environment on republican citizens. "If we see these things in the green tree, what may we expect in the dry?" asked a correspondent about the young factory laborer. Reuben Haines, a philanthropic manufacturer from Germantown, lamented the system of child labor in neighboring Manayunk after viewing Lowell and its "beautifully arranged" manufactories, decent boarding houses, and the "very lady-like females" who worked and lived in the New England mill town. "There appear none of the evils," he wrote from Lowell in late 1830, "we have been threatened with by large manufactories [in the Philadelphia area]."[47]

By the early 1830s, labor radicals were embellishing anticapitalist editorials with critiques of the evils of child labor as well as "the oppressive length of time we are forced to work." Workers in Philadelphia mills, like New England's family mills, labored an average of seventy-two hours per week, longer than those in the Manchester or Glasgow mills. Owing to the Factory Regulation Act, British mill workers toiled no more than ten hours a day after 1833, while their

Manayunk counterparts worked no less than twelve and sometimes thirteen hours.[48] In 1833 the "Working People of Manayunk" turned out over a wage reduction imposed by Joseph Ripka and bitterly complained that they were obliged to work

> thirteen hours of hard labor, at an unhealthy employment, where we never feel a refreshing breeze to cool us, over-heated and suffocated as we are, and where we never behold the sun but through a window, and an atmosphere thick with the dust and small particles of cotton, which we are constantly inhaling to the destruction of our health, our appetite and strength.
>
> Often we do feel ourselfs [sic] so weak as to be scarcely able to perform our work, on account of the over-strained time we are obliged to labor through the long and sultry days of summer.[49]

Concerned that such conditions were turning Pennsylvania's cotton mills into seats of discontent, the state legislature in 1837 launched an investigation into "the system of labor adopted in cotton and other factories." In its report to the Pennsylvania Senate, the Peltz Committee recommended prohibiting the employment of children under ten years and limiting the hours of children under sixteen to ten hours per day. "It is far better," the committee reported, "that we should forego pecuniary advantage rather than permit large masses of children to become the miserable victims of an oppressive system."[50] While the senate committee was concerned that the Pennsylvania textile manufacturers might be unable to compete effectively with those in other states if they regulated hours of labor, their greater concern was that the state's factories were producing an ignorant, immoral, and potentially dangerous class of citizens. Whatever the accuracy of the committee's assessment, the testimony presented at the hearings painted a stark portrait of the factory system of labor in the urban seaport.

While witnesses from Manayunk and Philadelphia mills singled out the long hours of work—seldom less than eleven hours per day—as the greatest evil of the factory system of labor, they did not limit their complaints to the length of the working day. Time after time, they pointed to the lack of uniformity or regularity of the hours of labor as a condition that allowed the employer to take advantage of the mill employees. "There is nothing which deserves the name of a system in the time of labor in the factories of Pennsylvania," the Peltz Committee concluded.[51] Work days at the Philadelphia mills varied from eleven to fourteen hours. Even the work schedule of an individual mill varied from day to day.

The timetables of the mills were adjusted regularly each season because of the varying amount of daylight. Mill employees commonly labored in the summer from sunrise until "the light is insufficient to see in the evening." With an hour and a half allowed for two meals, the typical summer schedule amounted to twelve-and-a-half hours of actual labor. When the days grew short in the winter months, the mill owners lighted the factories, which allowed work to continue on into the evening, typically until 8:00 P.M.[52]

Within this seasonally shifting pattern of work, mill owners made individual adjustments, extending the end of the winter working day, cutting into meal time, or ringing the work bell early in order to get more labor time from their operatives, who were paid a daily wage. Charles Kelly, for example, ran his Blockley mill in the winter from earliest light to 9:00 P.M., with only sixty-five minutes allowed for meals. Joseph Fleming of Fairmount required his mill workers to labor as much as thirteen hours and forty minutes during some winter days.[53] Witnesses before the Peltz Committee pointed to a variety of reasons why mill owners did not standardize the work day. The time of labor was often lengthened to make up for losses due to short days, repair days, and freshets that would wash away water wheels and power shafts or ruin canal water levels. The main cause for the varied work days, however, seemed to be competition among the textile manufacturers. Charles Hagner testified before the committee about how mill owners vied to increase output: "The additional labor of one hour in establishments, where from one to five hundred hands are employed, is an important item and offers great temptations to the employer to overwork his hands."[54] John Thornily, after nine months in the Manayunk mills, similarly suggested that the "want of uniformity in the hours of labor" was an exploitative tactic of profit-conscious manufacturers. He argued that the practice of some factories of calculating by the calendar and some by the "lunar months" confused families who contracted to work in the mills: "They do not know what they have bargained for; they do not know how many hours make a working day, or how many a week."[55]

In combination, the long work day, the unhealthy atmosphere of the cotton mills, the dangerous conditions of machine production, and the difficult labor processes made the early factory system a severe form of human labor. Both adults and children were subject to these conditions, but the exploitation of juvenile labor was particularly acute. Witnesses from the Philadelphia mills denounced factory work as unremitting and confining. The children, some as young as nine years old, were required to stay on their feet continually for twelve or more hours at a time. Children between the ages of nine and eleven labored at one of the most strenuous jobs, that of carrying the sixteen-pound boxes of bobbins up one to four stories between the preparation and spinning rooms. Older boys worked as piecers for the adult mule spinners or assisted the adult carders. Edward Makin, who worked in William Almond's Philadelphia mill, testified that mule piecers were "obliged to walk a great deal, and rapidly." The piecers and other children in the factory "have no time for sitting during working hours," another Almond employee complained.[56] Particularly injurious to the young female workers were the tasks of guiding the rovings through the drawing frames and tending the spinning throstles. This type of labor, the manager of a Blockley mill explained, required "constant stooping, the frames being about two feet from the floor."[57] Fatigued by the long, constant hours of physical labor, children would often fall asleep at their jobs and had to be strapped to be kept awake.

Although a reference to factories as "hells upon earth" had first been made with regard to the mills of Manchester, it applied equally well to the early mills of Philadelphia. Working in a dust-filled atmosphere in overheated rooms and standing and stooping for hours, Philadelphia workers developed serious disorders and diseases. The ankles of children and young adults swelled from the hours of standing, and children often complained of headaches. Serious lung diseases, among them a type of bronchial inflammation known as "spinners phthisis," were common in the unventilated cotton mills of Manayunk. A Philadelphia physician testified that he had treated many employees for scrofulous disease, a tuberculosis of the lymphatic glands. Adults as well as children developed stomach problems. The physical and mental trauma of labor in the mills was evidenced by the fact that female workers' menstruation cycles would frequently cease or be prolonged.[58]

Tending or working around the textile machines was also simply dangerous labor. Exposed gears and parts of machines frequently injured or killed children and adults. Charles Hagner had seen two children die when they were caught up in the mill machinery.[59]

For workers who had labored in English textile mills to testify that the factory system in Philadelphia was more oppressive than in England was perhaps the most telling indictment of factory conditions in the Mid-Atlantic and of a self-regulating market philosophy.[60] By the mid-1830s, English workers and reformers had gained a parliamentary act to correct some of the worst abuses of the factory labor system, but such conditions still existed in the mills of Manayunk. Ventilating machines, for example, which were extensively used in England to clear cotton dust from production rooms, were used in few mills in the Philadelphia area. Only a handful of mill owners in the region paid attention to guarding the gearing, which had become a common practice in British mills. Nor had Pennsylvania enacted protective factory regulations, as had England, that prohibited the labor of children under nine and set a maximum forty-eight-hour work week for older children. "What has been done for our children?" asked the working people of Manayunk, pointing to the fact that the "deplorable conditions" of child labor in English mills had been somewhat alleviated.[61] Agreeing with the testimony of Philadelphia workers, doctors, and some manufacturers, the senate committee issued a report that called for legislation to reform the system of child labor in the state.

Low wages and patterns of wage payments kept those who labored in Philadelphia's textile mills struggling for subsistence. The wage rates of Philadelphia mill workers fell below those of southern New England and more closely matched the wages of factory workers in Manchester or Glasgow. Statistics of the earnings of each type of mill laborer in the 1830s support James Montgomery's observation that the object of American cotton manufacturers was "to pay their help just such wage as will be sufficient for them to remain at their work."[62] A child operative in the mills of Philadelphia took home at most a dollar per week, as did children under sixteen in Lancashire mills. Testimony during Pennsylvania's factory investigation in 1837 revealed that most children earned no more than seventy-five cents per week. A few witnesses gave evidence that the youngest children received as little as fifty cents for six days of labor. In his denunciation of the system of child labor, Manayunk's school master swore that wages for children were even less:

> Under my own observation, some children of tender age have been dragged through the streets early and late, fair or foul, week after week, and month after month, for wages which would not amount to twenty dollars per annum. I state the wage as stated to me by the parents—seventy-five cents every two weeks![63]

Women in their late teens and early twenties, who constituted almost half of the work force in the Philadelphia mills, earned wages comparable to those of Lancashire females but less than those of the Lowell women in the mid-1830s. Throstle spinners in Philadelphia, for example, earned a maximum rate of $1.92 per week, the low end of the scale for female spinners in English mills. Lowell spinners, on the other hand, earned approximately $2.50 per week. According to a recent study, the female spinners in the Hamilton Company of Lowell received $3.48 per week in 1836. Only the Philadelphia powerloom weaver, the highest-paid female operative in the mill, fared better than her Lancashire counterpart, earning between $2.00 and $3.00 per week. Yet even she lagged behind the Lowell weaver, who averaged $3.15 per week in 1835. If a family of mill workers combined their weekly earnings, including that of the father who might earn as much as $6.00 to $7.50 as a mule spinner, they still had an income that was "barely sufficient to supply [them] with the necessaries of life."[64]

The boarding charges of the mill owners further reduced the meager earnings of their wage laborers. Cotton manufacturers in Philadelphia and Manayunk generally provided housing for their mill workers that took about 50 percent of their wages. In the early 1830s, Manayunk mill owners charged $1.00 to $1.25 per week for female lodgers, the same rate that Lowell operatives paid for both board and room. A Lowell employee was left with $1.90, on average, after paying for lodging, whereas a Manayunk powerloom weaver had $1.00 left to buy food and fuel for the week.[65]

Philadelphia mill owners further lowered their costs by occasionally paying mill workers in depreciated notes, which reduced the value of the laborer's earnings by 10 percent. Even the employers' pattern of wage payments worked to the detriment of the wage earner. While pay periods varied from mill to mill, the standard interval was four weeks. This extended pay period created difficulties for factory workers living close to the subsistence line, especially at the end of the month. Eighty-year-old Samuel Ogden, a former mill worker, told the senate committee of 1837 that the practice of "not paying the work people oftener than every four weeks, or calendar month, and keeping a few days, or one or two weeks' wages unpaid" was an "absolute evil."[66]

Periodic interruptions in production, beyond the control of either employer or employee, also denied the mill workers steady earnings throughout the year. Mill employees frequently lost days and often weeks of employment because of destructive freshets, lack of water for power, and routine machine and race repairs.[67] In 1831, for example, Borie complained to the Schuylkill Navigation Company that mill owner and operative alike were suffering because the company was negligent in restoring adequate water levels in the canal. All the mills in Manayunk were stopped for three days in 1838 for want of waterpower.[68]

In the *Germantown Telegraph* of September 18, 1833, the "Jeffersonian Workingman" explained how a mill worker's annual expected earnings easily could be reduced by lost weeks of work:

> We will see how the year will average—during which the canal company takes two weeks for repairs; back water by freshets in

the river, two weeks; holidays—Easter, Christmas, 4th of July, day to go and vote for the tariff, want of bobbins, machinery repairs, & c. two weeks more.

Thus a Manayunk mule spinner, according to the calculations of Jeffersonian Workingman, would lose six weeks of work per year and have an average weekly income of $3.94. Such irregularities of employment inherent in the early industrial system combined with low wages and relatively high boarding costs to keep some families dependent on mill work close to "poverty and pauperism," as Manayunk workers argued. It seems unlikely that the experience of those in the Manayunk and Philadelphia mills matched that of Rockdale's mill employees, who supposedly were able "to live well and save money."[69]

To enhance their control over their immigrant labor force and reduce costs, the Manayunk and Philadelphia mill owners instituted a system of rules thought to be essential in "all well-regulated mills." Printed and posted in identical form throughout the city's textile factories, mill owners used the nine general rules to discipline and regulate their labor force. This type of collective cooperation was an important step taken by early industrial capitalists in the struggle for control in the workplace. A. P. Usher has described such uniform sets of work rules as "symbolic of the new industrial relationship. Without any change in the general character of the wage contract, the employer acquired new powers which were of great social significance."[70] The fact that they became "a frequent bone of contention between employers and employed" in the Manayunk mills compels historians of the industrialization process to scrutinize their form and function.[71]

Both apologists and antagonists agreed that these rules were designed in the nature of a contract to be absolute and legally binding. "Engaging to work in a factory," the Peltz Committee concluded, "is considered as yielding assent to the laws." However, such a contract, regarded by capitalists as indispensable for the good management of their mills, was not one of mutual consent: "Necessity is the binding force," argued one labor activist.[72]

The content of the rules related "principally to the strict confinement of the hands to labor." Mill employers intended that their employees labor the maximum amount of time. Stiff fines were imposed for tardiness or absence from work without reason or notice: one-quarter of a day's wages for being fifteen minutes late; double the value of wages for absence without reason; and forfeiture of all past wages for leaving the mill's employment without notice. Two of the rules prohibited casual interaction between the workers: no one was allowed in the mill without permission; no more than one person was allowed outside one of the production rooms at the same time. And two regulations imposed stiff fines for harming the machines and for negligent or "badly done" work. "The necessity of the employer to retain this power [of discharge for such conduct] is obvious," Charles Hagner testified. The need to enforce these rules called forth the role of the mill manager and boss spinner, who together "direct[ed] all matters relating to the police and conduct of the people employed."[73]

The extent to which the mill operatives were willing to submit to the fines for either improper conduct or bad work is unclear. Many workers and some foremen and managers recognized these rules as an abusive use of power by the mill owners. Employers enforced them in a vigorous manner, usually without consulting the workmen. Mill workers found most contemptuous the rule requiring an employee to give two weeks' notice before quitting. In order to enforce this agreement, the mill owner retained "at all times" two weeks' worth of wages for each hand. Samuel Ogden, the retired mill worker, condemned the employers who enforced this practice because they routinely turned the tables and discharged hands without giving them notice. Ogden had good reason to know, since in 1834 Joseph Ripka had abruptly discharged his four daughters who worked as powerloom weavers. Although Lowell mill owners also instituted regulations that required workers to give two weeks' notice, these requirements were rarely enforced because, as Thomas Dublin suggests, labor was so scarce.[74]

Mill operatives could not necessarily find recourse in the courts against abusive actions by their employers. When Mary Ann Schofield left Joseph Ripka's mill without notice, Ripka withheld "the whole of the wages due her." Although Schofield sued Ripka and won a judgment against him, he simply delayed complying with the ruling. Two weeks later, he still had not paid Schofield what he owed her. "The fact is made well known," testified Ogden, "that the work people employed in factories seldom obtain their withheld wages from their arbitrary employers by law means."[75] The law, in the form of legal contract, operated primarily in the interest of the textile capitalists.

The manufacturers of Rhode Island, like those in the Mid-Atlantic port, also used poor families in their mills. However, both Rhode Island and Massachusetts mill owners during the 1830s had to recruit them from a relatively scarce labor pool in the rural hinterland.[76] This may be why the "fatherly" proprietors who owned the country mills of southern New England operated their mills in a paternalistic way, while Ripka, Borie, and other Philadelphians did not. John Amory Lowell, the son of Lowell's founder, stated as much, pointing out that in order to attract young women to their mills the early Lowell capitalists could not create a notoriously oppressive factory system. "Humanitarian inclination and the requirements of labor supply," Stanley Lebergott has written of the pre-1840 New England textile industry, "went hand in hand."[77]

The opening of the worldwide cloth market and the technological improvements of the powerloom and spinning throstle provided ample incentive and opportunities for Manayunk textile capitalists like Borie and Ripka to utilize an unskilled female and child labor force. And it was the labor force, its supply and skills, that was central in the evolution of capitalist productive relations in early industrial Philadelphia. Not dependent upon the skills of the independent and scarce tradesman as Nicholson had been, nor subject to the schedule of the outwork weaver or spinner like the Guardians of the Poor, Philadelphia's early industrial capitalists relied daily upon hundreds of unskilled women and children to attend to and labor at the pace of their highly capitalized machines. At the base of John Nicholson's early failure had been the erratic transition of the craftsmen and merchants into the system of capitalist manufacture. Traditional roles within the urban workhouses also had mediated the relations between the Guardians of the Poor and their workers. The relations of production of the mechanized mill, however, to a great extent now clarified the division and antagonisms between a laboring and a capital class in the old commercial manufacturing center. And indeed, in the early 1830s, the textile mill workers of the Manchester of America would organize to oppose the conditions of their labor and to challenge the prerogatives of the mill owners who created the industrial factory system in ways that were remote from the protests of Nicholson's glass workers or framesmiths. The industrial strikes by Manayunk textile workers against the oppression under which they labored marked the emergence of a militant labor movement in Philadelphia in the early 1830s at a time when operatives in the New England factory system were quiet.

IV

Economy and Society in Roxborough and Manayunk

Roxborough was almost a century-and-a-half old when it became the early home of the Mid-Atlantic's textile industry. Before the cotton factories crowded onto the Manayunk canal, a thriving flour industry lay at the base of work and wealth in the preindustrial township. Manayunk's early industrialists did not set their textile mills on the empty banks of the Schuylkill River; nor did the factory operatives flock into a sparsely settled township. A population of craftsmen, manufacturers, laborers, proprietors, and farmers whose productive lives and relationships had been tied to a vigorous market economy for a generation confronted the new industrial laborers and capitalists in the early nineteenth century. The social and economic order in which they lived conditioned their perceptions and responses to the factory and to its workers and owners.

In the eighteenth as in the nineteenth century, waterpower shaped Roxborough's development into a regional manufacturing center. One of the twelve original townships of Philadelphia County, Roxborough lay a half-dozen miles up the Schuylkill River from the docks of the Delaware. Within the first two decades of Pennsylvania's settlement, German and Quaker families bought sections of the original tracts of the township and established mill seats along the Wissahickon Creek, which flowed down the northeastern length of the township before cutting across the community to the Schuylkill. By the end of the eighteenth century, the small but rapid Wissahickon provided power for some of the largest flour mills in the Mid-Atlantic region.

Roxborough's eighteenth-century millers had chosen an advantageous site for obtaining raw materials and markets. They located their processing businesses in the heart of the country's largest wheat-producing region and only a few miles from Philadelphia, the most important flour entrepôt of the eighteenth century. By the 1750s, the international demand for Pennsylvania foodstuffs from England, Portugal, and the West Indies had made flour milling the most important manufacturing industry of the colony. The shipment of hinterland flour dominated Philadelphia's export trade throughout the rest of the century.[1]

During the 1790s, the Wissahickon and Brandywine creeks became the processing centers of the Mid-Atlantic flour trade. In 1795, Rochefoucauld-Liancourt counted eight Roxborough flour mills along the Wissahickon. Before the turn of the century, the two waterways were "noted for the best, and most numerous grist mills, in either this province or any other part of British America."[2]

In the postrevolutionary heyday of the export trade, commercial activity swirled around these mills. Farmers as far as seventy miles away knew the Wissahickon mills as a ready market for their grain. At times, the Ridge Road and mill roads that linked this turnpike with the Wissahickon would be lined for miles with wagons filled with wheat and rye. The innkeepers and tavernkeepers no doubt welcomed the harvest season when their "otherwise quiet township was enlivened by crowds of farmers and teamsters."[3]

The merchant millers of the Wissahickon competed aggressively for the grain and the flour trade. They ventured, according to one local historian, "continuously out on the roads day and night" to intercept the teams making their way down the Ridge Road to the mills.[4] Some millers traveled into the country and contracted with the growers for their harvests. By the 1790s the wealthiest mill owners, Jonathan Robeson and Joseph and Jonathan Livezey, bought grain from as far away as New York and Virginia. They invested in teams of oxen and horses to haul the grain from the docks in Philadelphia to Roxborough and to carry the barrels of flour back to the port. In the late 1780s and 1790s, millers constructed warehouses along the Schuylkill in the township itself. Here they could store the grain transported by longboat from the fertile farm lands of the upper Schuylkill.[5]

The postrevolutionary technological innovations of Oliver Evans allowed Roxborough merchant millers to increase production for the commercial market. In 1783, Evans conceived a system of mechanical devices that processed flour in a continuous and uninterrupted operation. His system of machines and conveyer belts, he claimed, lessened "the labour and expense of attendance of flour mills, fully one half."[6] In addition, entrepreneurs who invested in Evan's mechanical mill could realize profitable gains in output. This increase in productivity, one historian of America's flour industry declared, signaled the "transformation of the industry to a capitalist basis."[7]

Using the Evans system, Jonathan Robeson built up one of the largest milling businesses in the Philadelphia region. His "manufactory mill" awed his French guest Rochefoucauld-Liancourt. "It never ceases working", recorded Liancourt upon his visit in 1795. Robeson imported forty-five to fifty thousand bushels of corn from Virginia and New York that year to keep his mill constantly producing for the lucrative market.[8]

The tax assessment lists for the township indicate that mechanized milling facilitated the concentration of the flour industry in the hands of the big producers. To install the Evans system required a large amount of capital for the machinery and waterwheels capable of harnessing additional waterpower. Such an investment was no doubt responsible for causing the assessed value of the mills of the wealthiest millers of 1791 to triple within a decade. As the large millers consolidated control of flour production, they rose into the community's economic elite—those who were assessed for $2500 or more in real property. The substantial increase in productivity allowed by the Evans system evidently pushed the small producers out of competition, for the number of mills declined during the early nineteenth century, from eight in 1800 to five in 1819.[9]

The entrepreneurial millers who controlled Roxborough's lucrative flour industry belonged to the community's oldest families. The large landowners and custom millers of the mid eighteenth century—the Rittenhouses, Livezeys, and Robesons—became the large merchant millers of the Wissahickon flour industry who thrived in the international trade of the postrevolutionary era. These Quaker and German families had bought some of the first titles to Roxborough tracts in the late seventeenth century, and their descendants continued to appropriate wealth and property through intermarriage, inheritance, and land investments.[10]

Among the most economically powerful of the mill region families were the Rittenhouses. Wilhelm Rittenhouse arrived in Roxborough in 1690 and built one of the first paper mills in America on Paper Mill Run, a small rivulet that emptied into the Wissahickon. This enterprise passed from father to son for the following 140 years. Only one line of the family, however, followed the papermaking trade. In the mid eighteenth century, Nicholas and Henry Rittenhouse constructed the first two grist mills in Roxborough along the Wissahickon. By the Revolution, seven Rittenhouses owned one-tenth of the assessed acreage in Roxborough, including most of the land fronting the western bank of the Wissahickon in the lower half of the township. In 1782, the Rittenhouses added to their holdings by acquiring the century-old Holgate fulling mill and converting it to the more profitable production of flour.[11]

Investment in mills and mill property along the Wissahickon placed the Rittenhouses in a position to control a large portion of Roxborough's postrevolutionary flour trade. By the turn of the century, Martin, Abraham, and William Rittenhouse owned three of the eight valuable commercial mills. With two of their kin also operating paper mills, the family controlled a large portion of the Wissahickon mill seats in the township.[12]

Two Quaker families denied the Rittenhouses complete domination of the flour-processing business in Roxborough. Peter Robeson, like the Rittenhouses, was a descendent of one of the earliest settlers and grist mill builders along the Wissahickon. At the mouth of this river, which ran through his 500-acre township tract, Andrew Robeson constructed a grist mill in 1691. By the mid eighteenth century, the Wissahickon powered three Robeson mills. Through the end of the century, Peter Robeson controlled twice as much property and wealth as any Rittenhouse, and over the years his flour earned the reputation as one of the finest brands in the nation.[13]

The brothers John and Joseph Livezey, like their brother-in-law Robeson, turned their family's successful eighteenth-century mill into a large commercial enterprise after the postrevolutionary depression. Their father, Thomas, started in the milling industry in the mid eighteenth century. Thomas Livezey exemplified the successful merchant miller who competed in the prosperous flour export business. He maintained an extra team of horses always ready to meet the loaded grain wagons and assist them up the Ridge Road. To ensure his success in competing for local harvests, he sent John and Joseph out on the Lancaster Pike and other trade routes to "bargain for incoming loads of wheat and divert them to his mill." Like Robeson, the Livezeys delivered their flour directly to ships in Philadelphia bound for West Indian or European ports.[14]

In 1790, John and Joseph inherited one of the largest estates in Roxborough and one of the most lucrative milling businesses in the country. Only their brother-in-law owned a more valuable estate and mill. They expanded their investment quickly. John bought the century-old Gorgas mill in 1792, and when the original Livezey mill burned down in 1793, the brothers rebuilt and enlarged the three-story structure, converting it to the Evans system. In the same year, they built a warehouse in Philadelphia to handle their overseas trade. John Livezey bought a third Wissahickon mill seat in 1802, and by 1808 the Livezeys had added their own coopers' shops next to their Glenfern Estate mill.[15]

These Wissahickon capitalists, who coupled a propertied base acquired from their merchant-milling forefathers with entrepreneurial acumen, had sufficient assets to endure the downturns in the international market between 1816 and 1820, when the manufacture of grain was a "very unprofitable business."[16] While enlarging the eighteenth-century grist mills that had been in their families for decades, the Robesons, Livezeys, and Rittenhouses competed aggressively for grain crops, acquired warehouses, built cooper shops, and supervised the overseas marketing of their flour. Yet as their investments over the years pushed them to the top of their township's wealth structure, the old milling elite found their economic prominence within the community challenged by the largest landowning family in Roxborough.

The Leverings of turn-of-the-century Roxborough were also descended from one of the township's original settlers.[17] In 1691, Wigard Levering purchased one of the original tracts in the township. His sons and grandsons owned farms, trades, and inns in the community. Their wealth, however, was primarily based on the premium land Wigard had acquired between the Schuylkill and Wissahickon rivers. Blacksmith William Levering was the largest landholder in Roxborough on the eve of the Revolution. And in 1781, only three millers were assessed for more landed wealth than the tavern keeper Nathan Levering.

Whereas the economic interests of the Rittenhouse family lay along the Wissahickon and in Germantown, the Levering's were dispersed along the Schuylkill River. Through the turn of the century, as the millers of the Wissahickon built transportation links in the eastern part of the township, the Leverings developed the area along both banks of the Schuylkill. In 1791, Nathan Levering petitioned for a road from his "plantation on the Schuylkill" to the Ridge Road "whereby a communication may be had to either Church, Mill or Market."[18] The farming population and community's tradesmen settled on the Levering tracts along the Ridge Road. Here Nathan Levering rented out six of his seven dwellings to various craftsmen. On the eve of the arrival of the textile mills, only Robeson and the Livezey brothers were assessed for more wealth than Nathan Levering, and only Robeson came close to owing as much land as Levering in Roxborough.[19]

Before 1820, the work lives and opportunities of Roxborough's middling and lower classes were shaped by the township's emergence as a flour and marketing center. A snapshot of Roxborough at any time between 1790 and 1820, when the population rose from 778 to 1,682, portrays a far from equal society in which half to three-fifths of the taxable males owned no real property and more than half of the community's resources were held by the top 10 percent of the population (tables 1 and 2).[20] A more precise understanding of the direction of change in the structures of property holding and production can be obtained by comparing the levels of wealth at which different types

TABLE 1
Distribution of Taxable Real Wealth in Roxborough, 1791–1822
(Percentage of taxable wealth in each decile group)

Decile Group	1791 N = 156	1800[a] N = 185	1809 N = 260	1819 N = 281	1822 N = 319
Poorest 0–30	3.6	12.0	4.7	4.2	5.0
Middle 31–60	9.0	21.2	10.1	8.4	8.2
Upper 61–90	35.6	33.2	30.7	32.9	31.1
Wealthiest 91–100	51.8	33.6	54.5	54.7	55.7

Source: Philadelphia County Tax Assessor Ledgers, Roxborough, 1791, 1800, 1809; Tax Duplicate, Roxborough 1819, 1822; Archives of the City and County of Philadelphia, City Hall Annex, Philadelphia.

[a]The notable exception to the pattern of distribution appears in 1800 when the top decile controlled just one-third of the wealth. This deviation can be attributed not to a leveling of wealth but to the calculation of wealth distribution on an exceptionally low rate of tax on real property. The tax rate in 1800 was seven cents on every hundred dollars, much below the twenty- and fifty-cent rates of the other years. Since the personal rate on occupation is comparable throughout the thirty-year period, those people without real property were paying a higher proportion of taxes in 1800 than at any other time. Since these wealth distribution figures were calculated on the amount of tax paid on real and personal property, the 1800 figures reflect a regressive tax policy and not a more equitable distribution of wealth.

TABLE 2
Distribution of Population in Wealth Categories
(categories in dollars)

	No Property		1–499		500–999		1,000–2,499		2,500–		Total
	N	%	N	%	N	%	N	%	N	%	Number[a]
1791[b]	86	55.1	35	25.4	17	10.4	12	7.7	8	5.1	156
1800	101	54.6	26	14.1	27	14.6	17	9.2	14	7.6	185
1809	146	57.0	43	16.8	34	13.3	19	7.4	15	5.8	256
1819	148	52.7	17	6.1	32	11.4	38	13.5	46	16.4	281
1822	191	59.9	28	8.8	33	10.4	37	11.6	30	9.4	319

Source: Philadelphia County Tax Assessor Ledgers, Roxborough, 1791, 1800, 1809; Tax Duplicate, Roxborough, 1819, 1822; Archives of the City and County of Philadelphia, City Hall Annex, Philadelphia.

[a]Widows and sons assessed a per head tax without an occupation have been excluded in the calculations.

[b]1774 figures are not included because a different base index would apply. In 1774, 47.6 percent of the taxables had no property.

of producers lived between 1790 and 1820. It was in this era that subtle yet significant changes began to differentiate the experience and interests of the community's producers.

Between 1790 and 1820, farmers were losing their niche in Roxborough's prefactory economy. There were few farm families left in the township to witness the coming of the factories after 1820. The rise of commercial milling in the area seemed to have had the dual effect of driving up the cost and reducing the supply of land. Bruce Laurie has argued that by the last third of the eighteenth century, farmers' sons with small holdings were turning to tenancy or nonagricultural trades in the Philadelphia area.[21] This seems to have been the case in Roxborough as commercial millers and large landholders invested in the township's acreage along the Wissahickon and Schuylkill.

In the waning years of the Revolution, one-third of Roxborough's ratables were farmers, 60 percent of whom owned the land they worked. In 1791, Roxborough's farmers made up 29 percent of the taxable inhabitants and over half of them were tenants. Another 27 percent of those who identified themselves as farmers subsisted on much less than the seventy-five acres that James Lemon argues were required to grow a surplus in southeastern Pennsylvania. Thus more than 80 percent of all farmers in Roxborough worked small acreages of land as tenants or subsistence farmers (tables 3 and 4).[22]

By 1800, as table 3 indicates, the number of farmers had declined to one-fifth of the taxable population. During the 1790s, when the flour-milling industry expanded, twenty-five of forty-five (56 percent) of the Roxborough farmers on the 1791 tax list disappeared, all but two of whom were tenant or subsistence farmers. The number of subsistence farmers, those cultivating less than seventy-five acres, dropped slightly. Much of the land formerly worked by small farmers was acquired by the millers and Nathan Levering, all of whom substantially increased their property holdings over the decade.

The rising price of land in the township is one factor that explains why tenant and subsistence farmers declined in Roxborough at the same time that wealthy millers, the Leverings, and a few large landholding farmers acquired more acres. According to Rochefoucauld-Liancourt, land in the area doubled in price just between 1791 and 1795, largely reflecting the start of construction by the Schuylkill and Delaware Navigation Company on the canal in 1793. The rising value of farmland between 1791 and 1800 pushed all classes of propertied farmers into higher wealth-holding categories, even though they added nothing to their acreage. But more important, it priced prospective farmers out of the land market. Only four out of the forty-five prop-

TABLE 3
Distribution of Occupational Groups in Roxborough, 1791–1828

Occupational Group	1791		1800		1809		1819		1828	
	N	%	N	%	N	%	N	%	N	%
Unskilled laborers	43	27.5	30	16.2	43	16.8	50	17.8	248	40.6
Farmers	45	28.8	37	20.0	43	16.8	46	16.4	59	9.7
Retail crafts	38	24.4	65	35.1	103	40.2	115	40.9	164	26.8
Building crafts	8	5.1	18	9.7	18	7.1	20	7.1	64	10.5
Food processing crafts	3	2.1	14	7.6	15	5.8	13	4.6	16	2.6
Retail services	5	3.2	9	4.9	22	8.6	23	8.2	31	5.1
Professionals	1	0.6	3	1.6	3	1.1	4	1.4	15	2.1
Mill owners and manufacturers	13	8.3	9	4.9	9	3.5	10	3.6	14	2.3
Total	156		185		256		281		611	

Source: Philadelphia County Tax Assessor Ledgers, Roxborough, 1791, 1800, 1809, 1828; Tax Duplicate, 1819; Archives of the City and County of Philadelphia, City Hall Annex, Philadelphia.

ertyless farmers of 1791 who stayed in Roxborough had obtained land by the end of the century.[23]

Poor fertility and the high cost of farm labor, combined with the rising price of land in the 1790s, drove many small Roxborough farmers off the land. Subsistence agriculture became untenable on marginally productive yet expensive land, making it more sensible to sell than stay. "The land is on the whole, of very inferior quality in this district," wrote Rochefoucauld-Liancourt upon inspecting the farming efforts of Peter Robeson in 1795. Joshua Gilpin, who traveled through the area in 1809, noted in his journal that the soil in the area was "by no means rich." Coupled with the low productivity of the expensive farm land, the "scarcity and great expense of labourers" at the turn of the century thwarted the small farmer's ambitions of advancing into commercial farming. Only the merchant millers like Robeson were able to procure day labourers without much difficulty in spite of their "scarcity and expense."[24]

Tenant farming continued to decline in the early nineteenth century. Half of Roxborough's farmers of 1800 were gone by 1819, and two-thirds of those who survived were without property. The completion of the Ridge Turnpike to Philadelphia in 1812 and construc-

TABLE 4
Real Property Holding of Roxborough Farmers, 1791–1809

	No Property (Tenant Farmers)		Less than 75 Acres		75 Acres or more		Total
	N	%	N	%	N	%	
1791[a]	25	(56)[b]	12	(27)	8	(17)	45
1800	21	(56)	8	(22)	8	(22)	37
1809	19	(44)	19	(44)	5	(12)	43

Source: Philadelphia County Tax Assessor Ledgers, Roxborough, 1791, 1800, 1809; Archives of the City and County of Philadelphia, City Hall Annex, Philadelphia.
[a]The 1819 Tax Duplicate does not enumerate acreage so it has been excluded.
[b]If four farmers' sons are excluded, the percentage would drop to 47.

tion of the Schuylkill Canal beginning in 1815 further drove up the price of land surrounding these transportation links. For those who could invest in farmland in Roxborough, the healthy demand for grain due to the Napoleonic Wars and the War in 1812 made commercial farming a profitable livelihood. But by 1819, farmers composed only 16 percent of the taxable population. On the eve of industrialization, agriculture was left mostly in the hands of large commercial farmers.

In 1819, when farmers had been reduced to one-sixth of Roxborough's taxables, four-fifths of the community's men were tradesmen or laborers, many of whom found work in the paper, iron, saw, and flour mills powered by the Wissahickon (table 3). The paper mills, including the old Rittenhouse mill, also employed a number of women and children, as did a small, short-lived spinning mill. The mill workers concentrated in what became known as Rittenhousetown where the dwellings were occupied principally by the laborers in the mills along the Wissahickon.[25]

The fact that two-thirds of all taxables in Roxborough's trades and services in 1819 held no real property suggests that we look more closely at the nature of property attainment, mobility, and productive relations that were overcoming the processing economy. Roxborough's thriving mill-based economy of the 1790s brought in young carters, millers, millwrights, and coopers from outside the community while providing employment for the sons of the township's farmers and craftsmen. Yet those young men did not stay and obtain land in the community of their birth. Fifty-five percent of these taxables of 1791 were gone by the turn of the century. Over the two decades preceding the Panic of 1819, only young propertyless men beginning careers in

TABLE 5
Wealth of Roxborough Coopers

	No Property		1–499		500–949		1,000–2,499		2,500–		Total Number
	N	%	N	%	N	%	N	%	N	%	
1791	7	(44)	9	(56)	0	(0)	0	(0)	0	(0)	16
1800	15	(44)	12	(35)	4	(12)	3	(9)	0	(0)	34
1809	29	(55)	16	(30)	6	(11)	2	(2)	0	(0)	53
1819	36	(55)	4	(6)	15	(23)	7	(11)	4	(6)	66

Source: Philadelphia County Tax Assessor Ledgers, Roxborough, 1791, 1800, 1809; Tax Duplicate, 1819, Archives of the City and County of Philadelphia, City Hall Annex, Philadelphia.

the cooper's trade tended to stay in Roxborough.[26] By focusing on how patterns of real wealth holding changed among coopers, the dominant craft in the flour-manufacturing center, we can gain further insight into the changing world of production in preindustrial Roxborough.

In 1791, sixteen coopers plied their trade in Roxborough. A tenth of all taxables in the community, these skilled producers of barrels for the merchant millers included propertyless journeymen and a few small propertied masters. As shown in table 5, the wealthiest cooper had an assessed estate of less than $500. Only possession of a few acres separated the majority of small independent craftsmen from the rest of the coopers who were propertyless.

Over the next thirty years, involvement in a booming export market affected the scale of, as well as productive relations within, the coopers trade. As the demand for barrels expanded, a few established eighteenth-century coopers accumulated capital, invested in larger shops, and hired the labor of a growing number of wage-earning coopers. From Roxborough's tax lists we can glean evidence of the development of two classes of coopers: the capitalist entrepreneurs and the propertyless laborers whom they employed.[27]

As the number of coopers grew rapidly, reaching one-quarter of the taxable population in 1819, the division of wealth among them widened. In 1819, the majority of coopers in Roxborough held no property, while a dramatic decline occurred in the proportion of those who composed the smallest category of property holders ($1–499 in assessed wealth; see table 5). Among coopers who had taxable property in 1819, most were in the middle or upper-middle level of wealth holding, and for the first time, a few coopers appeared among the economic elite of the community. Those coopers, who were the first of their trade to enter the highest wealth-holding category in 1819, were

TABLE 6
Property Mobility of Roxborough Coopers

	1791–1800		1800–1809		1809–1819	
	N	%	N	%	N	%
Remain Propertyless	0	0	8	29	12	29
Obtain Property	4	33	4	14	12	29
Property Holding Remains the Same	3	25	13	46	2	05
Property Holding Increases	5	42	3	11	15	37

Source: Philadelphia County Tax Assessor Ledgers, Roxborough, 1791, 1800, 1809; Tax Duplicate, 1819, Archives of the City and County of Philadelphia City Hall Annex, Philadelphia.

men who had lived and owned property in Roxborough for at least thirty years. The two wealthiest coopers at the turn of the century, Lawrence Miller and Andrew Merwine, had worked at the cooper's trade in Roxborough since at least the Revolution. Among the small landholders in 1791, their property tripled in value by 1800.[28] In 1809, when the majority of coopers for the first time owned no real property, Miller and Merwine, along with nine other Roxborough coopers, owned their own shops. No cooper accumulated more wealth than the entrepreneurial Miller, who joined the merchant miller Peter Robeson and the landed Levering family members in the economic elite of the community in 1819.[29]

Longtime residence in the community did not automatically assure economic security or upward mobility. As outlined in the tax lists, the contrasting careers of coopers Jacob Everman and John Bigony with their peer Lawrence Miller illustrate this point. Everman had lived in the community since at least 1791, and Bigony had appeared with Miller on the 1783 tax list. In 1819, Everman held no assessable property, as had been the case for twenty-eight years. Bigony, who had acquired a shop and small amount of acreage in 1809, was propertyless again.

By 1820, propertylessness and transiency had become the common experience of Roxborough's most populated craft. Few of the propertyless coopers of 1800 ever acquired land in their community. Eleven of the fifteen craftsmen left the community or remained without property through the next two decades (tables 5 and 6). We can only speculate about the fate of those who disappeared from the township's tax rolls at the turn of the century, but it is known that this was a difficult time for wage earners. The downward trend in real wages, which began after the Peace of Amiens and continued with Jefferson's embargo in late 1807, undoubtedly reduced the incomes of Roxborough's coopers. Barrelmakers would not have found better opportunities in Philadelphia, where the rising cost of living was also outstripping wage rates and continued to do so until the War of 1812. Thus, though George Moose, a propertyless copper in turn-of-the-century Roxborough, left before 1809, he was back by 1819. He returned to his community as he had departed, a poor journeyman.[30]

It can be inferred from the evidence on real property accumulation and migration that capitalist productive relations had altered the traditional trades of prefactory Roxborough just as they had in neighboring Germantown in the late eighteenth century. "In many of the occupations," concludes Stephanie Wolf, "a wide variation between the average and the median value of an occupation, as well as placement on the tax lists, points to large shops with wealthy entrepreneurs and a considerable contract labor force working under them."[31] Thus a subtle but decisive process was underway in the processing and commercial economies of greater Philadelphia prior to industrialization. This change, whereby coopers, farmers, and millers experienced growing divisions in the control of production and property, would play its part in the varied responses to the coming of the factory system.

After 1820, new economic and productive relations transformed the world of the small producer and master manufacturer much more dramatically. As rapidly as the mills were constructed on the canal, cotton textiles superseded Wissahickon flour as the base of the township's and the region's economy. A core of early industrial capitalists ousted the merchant millers as the economic elite of the community, and hundreds of immigrant factory workers, predominantly women and children, supplanted coopers, papermakers, and millers as the most common producers in Roxborough.

The cotton textile mills and other water-powered factories that crowded onto the narrow strip of land between the canal and the Schuylkill River ran the length of a sparsely settled section of Roxborough meadowland that gave way to a rocky-tiered slope leading toward the flatlands of the Ridge Turnpike. Prior to completion of the Flat Rock Canal in 1819, a few structures and dwelling houses, owned mostly by members of the Levering family, dotted this western section of the township. They were concentrated around the Leverington Hotel near the Ridge Turnpike, high above the Schuylkill and the canal.

Within a decade of the first sales of waterpower in 1819, factories and row houses for working-class families fanned out from the canal, turning the meadowlands into the population center of Roxborough and the industrial center of the Delaware Valley. "Five years ago," John P. Binns told a Manayunk crowd in September 1827, "there were but three houses in this place, with probably twenty inhabitants; now it is one of the greatest manufacturing establishments in the vicinity of Philadelphia."[32] The rate of population growth attests to the phenomenal rise of an industrial center within the township. In the thirty years preceding 1820, the decadal increase in Roxborough's population varied between 20 and 34 percent. Between 1820 and 1830, however, the township's population increased 98 percent from 1,682 to 3,334 inhabitants. By 1830, almost half of Roxborough's residents lived within the boundaries of Manayunk, which included less than a tenth of the township's acreage. In the next decade, the population boom in Manayunk increased the size of the township to 5,797 residents. The manufacturing town itself increased its number of inhabitants by 151 percent between 1830 and 1840, at which time the borough of Manayunk was incorporated.[33]

A decade-long housing boom of Manayunk reflected the township's urbanization. In 1825 fifty houses were under construction in the area near the mills. Dwellings could not be built fast enough. "They are in immediate demand and often bespoke as the foundation is laid," Binns remarked in 1827. In 1830 inhabitants of the manufacturing town occupied 200 new houses. By 1834 the numbers of homes had doubled, and a resident, echoing Binns' observation seven years before, reported population increasing so rapidly that houses were frequently "engaged before the first floor or joists were laid."[34]

A good proportion of the demand for housing came from the growing ranks of mill workers. "The population of Manayunk consists almost wholly of persons employed in the factories," announced Isaac Baird, Manayunk's first cotton yarn manufacturer when he described Roxborough's new residents in 1825. By 1828, 785 persons, well over half of Manayunk's residents, were employed in the five large cotton mills on the canal. Half of Manayunk's population were children under fifteen years of age. "The reason for this is obvious," wool manufacturer Charles Hagner commented. "The employment offered to children in the manufacturing establishments, is an inducement for families . . . to settle in this village."[35] By 1828 local residents marveled at the transforming effect of mills and mill workers on their

---

Opposite, Manayunk and Roxborough Township, ca. 1835 (adapted by Matt McGrath from Charles Ellet, Jr., "A Map of the County of Philadelphia from Actual Survey Made under Direction of Charles Ellet, Jr., June 30, 1839," (Historical Society of Pennsylvania).

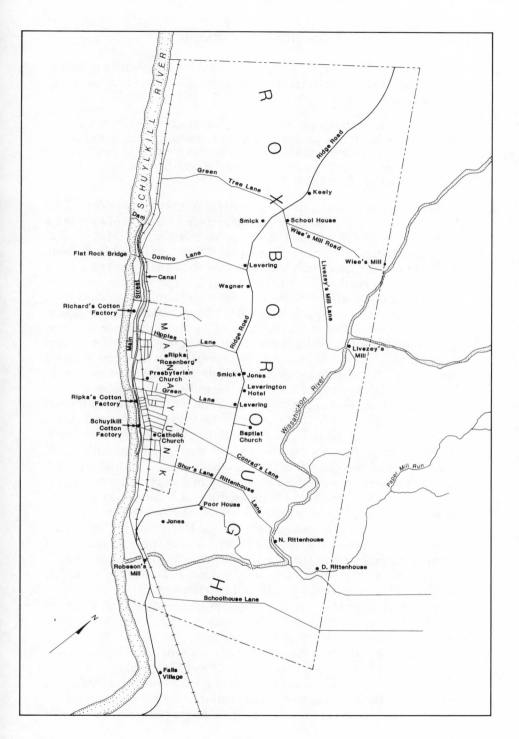

community. "The whole scene is changed," wrote William Young. The banks of the Schuylkill now lay under the "broad shadow of the cotton factory."

> A flourishing and populous village has risen up suddenly and where we but lately paused to survey the simple beauties of the landscape . . . the eye is arrested by the less romantic operations of a manufacturing community, and the ear filled with the noise of ten thousand spindles.[36]

The mill population grew at an even greater rate in the early 1830s than in the late 1820s. In 1832, 961 operatives worked in Manayunk's cotton mills, which accounted for approximately one-third of Roxborough Township's entire population. Over 100 individuals were employed in the two woolen factories on the canal. In 1834, a local correspondent made the same point that Charles Hagner had made six years before: the demand for child labor had brought 3,000 persons, consisting mostly of large families of moderate circumstances, to the township. On the eve of the Panic of 1837, seven large cotton mills gave employment to over 1,000 workers, most of whom were women and children. The families of mill workers lived in the neighborhood just beyond the canal, residing in tightly packed rows of family dwellings that ran three streets deep from the mills on the Schuylkill.[37]

Tax lists are silent about the women and children who composed approximately four-fifths of the cotton mills' labor force and three-quarters of the general factory population. The assessment roles, however, reveal that male mule spinners, machinists, and other mill operatives had become a relatively large proportion of Roxborough's taxable population, forming a new group of industrial laboring men within the township. In 1828, for example, 91 out of 642 males in Roxborough (14 percent) worked in the cotton and woolen mills on the canal. By 1835, approximately 215 of Roxborough's male residents, or 22 percent, listed their occupation as spinner, weaver, or manufacturer. By the early 1830s, well over 100 men in Roxborough found work in five other factories on the canal—one paper mill, a rolling mill, a saw-grinding mill, and two flour mills.[38]

The demand for housing and road improvements, as well as for the construction of factories and shops, spurred a great increase in Roxborough's population of construction tradesmen. In 1828, a Manayunk manufacturer announced that stonemasons, millwrights, and carpenters (along with machinemakers) "would all find plenty of employment here."[39] Indeed, the number of Roxborough's taxable males who earned their livelihoods in the building crafts more than tripled between 1819, the year the canal was finished, and 1828, when houses were in great demand. Unskilled laborers, according to a local booster in 1828, could also find good wages in Manayunk, and between 1819 and 1828, the number of laborers in the township doubled. Street improvements, bridge construction, and the building of the Philadelphia, Germantown, and Norristown Railroad to Manayunk in the early 1830s no doubt employed many young laborers in the area.[40]

The growing number of Manayunk factory employees helped to support fifty-seven more retail craftsmen and service tradesmen in the township in 1828 than in the previous decade. In that year, Andrew Young of Manayunk announced "a constant demand for mechanics of every description." In this boom period, $10,000 in factory wages were reportedly put into circulation in the community every four weeks. "A new and vigorous spring is given to everything," Young boasted. In 1837 Charles Hagner noted the particular economic interest that bound thousands of mill workers, tradesmen, and laborers of Roxborough's mill village: "Manayunk is exclusively a manufacturing town—the whole population subsisting, directly or indirectly through the manufactories established here."[41]

As the "vigorous spring" of industrial production increased jobs, commercial activity, and productive wealth in Roxborough, the proportion of the population who were propertyless grew. Rates of propertylessness and poverty in industrial Roxborough far exceeded those of the prefactory township. In any given year prior to the coming of the textile mills, approximately half of the male taxables held real property in the township of their residence. In 1841, the first year after 1822 for which a Roxborough tax list survives, two-thirds of the male taxables in the township were assessed as having no real property. "The poor hard working villagers" of Manayunk, in the words of one resident, filled out the ranks of the township's propertyless class. The largest single occupational group among those male residents identified as holding neither personal nor real property were 208 textile factory employees. These propertyless spinners, machine operatives, weavers, finishers, carders, rollers, and dyers composed 89 percent of Roxborough's 235 male factory operatives. In 1841, with their ranks decimated by five years of depression, this factory proletariat constituted 18 percent of the adult male population of Roxborough.[42]

At the same time, the propertyless rate within the traditional trades and crafts had soared with the advent of the industrial economy. In 1841, 86 percent of the retail craftsmen and 76 percent of the building tradesmen held no property compared to 52 and 40 percent respectively in 1819. Because many more women and children than men worked in the mills, we can assume that many had husbands and fathers who were among the township's nonfactory tradesmen and laborers. In 1840, only 94 of approximately 1,035 Roxborough households that contained members employed in manufacturing were headed by women. Most of the adult female mill workers, in other words, lived as boarders or were wives and daughters in households headed by tradesmen, laborers, or male factory operatives.[43]

Poverty, like propertylessness, increased with the coming of the factories to Roxborough. Before the period when "many poor families" sought work in the mills, the township, through the Directors of the Poor, dealt with its impoverished residents by boarding them with individual householders in the community. At a meeting of the taxable inhabitants early in 1833, it was decided that the "increasing number of poor rendered the system of boarding out impractical."[44] The growth of an industrial-based economy brought the need for a poorhouse. Roxborough raised the poor tax and used the revenue to build a poorhouse at the southern end of the township for its destitute inhabitants.

As propertylessness and poverty enshrouded the lives of a growing number of the township's inhabitants, the community's wealth became more concentrated. The tax lists reveal that in the years preceding the completion of the canal in 1819, about 55 percent of Roxborough's real wealth was controlled by the top tenth of the township's taxables. At the end of the 1830s, that elite held 65 percent of the taxable property. The wealthiest among them were the Manayunk textile mill owners in the township. Unlike the corporate mill owners of Lowell, who lived in Boston and appointed managers to run their mills, the majority of the cotton, woolen, and other factory owners on the Schuylkill relocated their residences from Philadelphia to Manayunk, near the mills they supervised. These early industrial capitalists were a diverse lot. Some had come up through the ranks of master manufacturers, some had immigrated to Philadelphia after the turn of the century; and some were successful merchants. As we saw in chapter 3, Joseph Ripka, an Austrian-born master-turned-manufacturer, and J. J. Borie, a French refugee and export merchant, owned the largest textile mills in Manayunk in the late 1820s and 1830s. Ripka moved to Manayunk in 1832 after building his third mill on the canal and a baronial estate he called "Rosenberg," located on the hills above the mills and tenements. Jerome Keating, the resident owner of the Borie, Laguerinne, and Keating Schuylkill Factory, lived in a mansion that stood at the end of the last row of his workers' housing. Although J. J. Borie did not move to the township, he bought a 130-acre estate on the Schuylkill within two miles of the textile town, where he lived in conspicuous elegance. After paying $22,000 for the site, he added a large stone barn and a row of forty stables.[45] The young brothers, Thomas and James Darrach, were only thirty and nineteen respectively when they took up residence in Manayunk to operate their woolen factories. Their father had been a successful hat manufacturer in Philadelphia. Brothers Samuel and Tobias Wagner were born in the port city in the 1790s and commenced cotton textile production in Manayunk in 1828 with $70,000 they had accumulated in a Philadelphia merchant and auctioneering firm.[46]

These individual industrialists formed an economically powerful sector of Roxborough's capitalist class. Their substantial investments enabled them to survive the Panic of 1837 and to maintain their position in the top 10 percent of wealth holders in 1840. Joseph Ripka, the Wagner brothers, the Darrach brothers, and James Kempton, who bought a large interest in the Schuylkill factory following the deaths of Keating in early 1833 and Borie in November 1834, were among the ten wealthiest property holders in Manayunk. Ripka was by far the wealthiest individual in the town, controlling more than twice as much real property as the next richest man. His four mills and estate were assessed at $80,000.[47]

Only three longtime inhabitants of Roxborough and the nearby Falls of Schuylkill became neighboring manufacturers of the newly arrived textile capitalists on the canal. One was Charles Hagner, who was born in Philadelphia and raised at the Falls, where his father owned a considerable amount of property and operated a paper and oil mill. The young Hagner took over his father's mills on Fall Creek in 1817, and when the canal was completed upriver, he moved his manufacturing business to Manayunk. Hagner was the second purchaser of waterpower on the canal, and in 1821 he built a drug and a fulling mill. The other proprietors on the canal with roots in the community were partners George Smick and Samuel Gorgas, who established a large flour mill in 1825. Smick's father owned and farmed a large tract of land in the township that ran from the Ridge Road to the Schuylkill River. Samuel Gorgas came from a long line of successful Wissahickon millers. These men owned and operated their mills through the 1830s. Compared to the textile manufacturers who arrived in the township after 1825, however, they remained small proprietors.[48]

While most of the owners and operators of the mechanized factories on the canal were newcomers to the township, a number of longtime residents profited from the development of the mill town itself. No one could have been more pleased with the ascendancy of Manayunk than the Levering family. When the Schuylkill Navigation Company began work on the canal in 1815, three Leverings were among the eight individuals who owned the land that comprised the future borough. Anthony Levering, who owned part of the meadowland on both banks of the canal, noted that "some benefit may arise from the canal on the Schuylkill River," when he wrote out his will and left each of his three sons two acres of land along the canal in 1818. And, indeed, in the year that the canal was opened, he sold five acres of adjacent land for $5,000.[49]

The Leverings of the factory period, like their forefathers, appeared on the tax lists as land-rich families of skilled craftsmen and farmers. The latest generation stayed in Roxborough, inheriting smaller but ever-more-valuable plots of their fathers' tracts. In 1832 seven Leverings appeared on the state tax rolls for the township, which catalogued the community's economic elite who possessed gigs, carriages, mortgages, stocks, and bonds. Peregrine Levering took over the family mantle of economic patriarch and public servant. In the family tradition, he learned a trade, that of carpentry, while accumulating capital in land development and contracting in Manayunk. Becoming a considerable landowner, by the end of the 1830s he controlled over $13,000 worth of real property in Roxborough.[50]

The wave of industrial and commercial expansion did not open new areas of investment for the Wissahickon millers as it did for the Leverings and other westside landholders. Only the Robesons, who were the largest merchant millers in the prefactory period and strategically located at the mouth of the Wissahickon, profited directly from industrial development. Peter Robeson saw his property values multiply after the canal was completed, and he sold some of his land in parcels to the cotton mill coowners J. J. Borie and Peter Laguerinee.[51]

The other merchant flour millers persisted as a wealthy enclave in the Wissahickon section of the township. The Livezeys, for example, oversaw 500 acres of land along the Wissahickon in 1833. Joseph Livezey and his son Thomas were assessed for more personal wealth, in the form of mortgages, bank stock, ground rent, furniture, and carriages, than any other taxpayer in 1832. Like the Livezeys, the Rittenhouses, conspicuous on the tax lists as a type of outlying gentry, were also isolated from the economic development of industrial Manayunk. They endured as a reminder of the old landed elite. Only two Rittenhouses operated mills after the 1820s. Nicholas Jr. and Martin carried on the family's only flour mill, and Enoch Rittenhouse converted one of the family mills to papermaking. Enoch and Nicholas Jr. were both assessed for large estates at the end of the fourth decade of the century. They were identified as gentlemen on the tax list of 1841; the other members of the Rittenhouse family were recorded as farmers.[52] As we shall see, the divisions of settlement and economic interest that separated these old merchant millers from the new industrial mill owners would partly determine the political battle lines drawn in the early industrial community.

Roxborough's industrial working class began to form with the influx of families seeking work in the mills. As indicated earlier, the recurring trade and market crises of the English and Irish industrial economies provided Philadelphia mill owners with laborers who needed, and were familiar with, factory employment. The extant sources (no payroll or factory account books of the Manayunk mills have survived for this period) tell us most about the adult male machine operatives in Manayunk and Roxborough. Of the 229 male operatives who appeared on Roxborough's 1835 tax list, only 34 (15 percent) had been recorded in the community in 1828 when manufacturing growth took off at a feverish pace. Most of them were weavers, mule spinners, machinists, and "manufacturers" who had carried on the same trade in the town in 1828. Only 15 unskilled men who lived and labored in Roxborough in that year—laborers, carters, and cordwainers—appeared on the tax list seven years later as textile factory workers. Thus the male textile labor force of 1835 was not composed of Roxborough's skilled and unskilled sons.

The immigrants who filled the ranks of the factory laborers and the village row houses quickly changed the cultural and demographic profile of Roxborough Township. "The whole village is a kind of theater, in which hundreds of people, composed of different countries, are grouped together," a local manufacturer wrote of Manayunk in 1828.[53] Records of the four Manayunk churches founded between 1827 and 1840 provide confirmation of this influx of immigrants into the mill community. Gravestone inscriptions indicate that the original members of both the Methodist and Presbyterian churches had emigrated from the English textile districts of Yorkshire and Lancashire. Twelve families from Ireland were among the twenty who founded Saint John the Baptist Catholic Church, and "people from Ireland" constituted the parish of the Manayunk church in the 1830s. The church's burial records confirm its Irish-American constituency. Between August 1832 and August 1835, 49 adults were buried in the church's cemetery; all but three had been born in Ireland.[54]

Through various sources, the immigration patterns and work experience of more than a dozen English, Irish, and German textile operatives and their families who worked in Manayunk between the late 1820s and 1840 can be pieced together. In each case, the mill workers had arrived in America between 1806 and 1836. All but two immigrated after 1820. All but four of these families and individuals came from England, mostly from Lancashire or Yorkshire. An Irish couple, Neal and Margaret Loughrey, brought their eight children to Manayunk in 1829 to work in the mills. A German couple named Rudolph and their seven-year-old son arrived in the textile town in 1836 and soon entered the mills.[55]

Although some male English immigrants headed alone for American textile districts, many arrived in Manayunk with their wives and children, or often, it appears, with a single son. Nine-year-old James White, for example, immigrated with his English parents in 1814. He appeared in the 1828 Roxborough tax list as a spinner, no doubt his father's trade. For the next nineteen years, he labored in the mills on the canal. Children commonly entered the mills upon reaching Philadelphia, where eleven-year-olds and younger constituted one-fifth of the textile factory labor force. Sebastian Rudolph began work as a piecer at age seven soon after his arrival with his parents. James Gartside was eight when he disembarked in Philadelphia with his father, a Lancashire weaver, and went straight to work in a Blockley mill as a piecer. When he moved to Manayunk in 1835, he entered a canal mill as a twelve-year-old with four years experience.

Many of the adults who immigrated to Manayunk or other American textile districts knew well the world of the factory workplace. Thomas Mosely, for example, arrived in America in 1824 at the age of twenty and appeared in Manayunk three years later. A native of Yorkshire, he had gone to work in a mill when he was nine. Jeremiah Wilkinson, who supervised one of Joseph Ripka's throstle rooms in the mid-1830s, had become, in his own words, stunted and lame from "early labor in the cotton factories in England." In 1837, by the time he was thirty, he had spent twenty years in cotton mills in England and America. A fellow operative of Wilkinson's after immigrating in 1832, Thomas Bolton testified that he had been "brought up in the cotton business." Samuel Ogden had arrived in one of America's coastal ports in 1806. He also labored for Ripka as a spinner. In 1837 he was nearly eighty years old and had spent thirty years working in cotton factories on both sides of the Atlantic. Luke Neld had spent twelve years in a Manchester mill before taking passage to America. Robert Craig arrived in Philadelphia and entered a Blockley mill at age twenty-three with thirteen years experience in Irish mills. The mill operatives of Manayunk, we may say, were members of a transatlantic industrial labor force, deriving their experience and sentiments as working people from a mobile life between British and American mill towns.[56]

Persistence rates among textile operatives once they arrived in Manayunk indicate that the mill worker moved often, though not so frequently as the rural daughters of New England who found jobs in the Lowell mills. Over a seven-year period from 1828 to 1835, only one out of five male mule spinners and manufacturers remained in the textile town on the Schuylkill.[57] The records of one Philadelphia cotton textile mill in the northeast section of the city, only a few miles east of Manayunk, shed further light on the mobility rates of female powerloom weavers. In 1829, ten different women appeared in William Whitaker's monthly accounts for the powerloom department. Each had worked an average of two-and-a-half years for Whitaker, more than twice the fourteen-month average persistence rate of their sisters in the Lowell mills. The experience of a sample cohort taken from the work rolls on a single day also indicates that urban textile workers in Pennsylvania persisted longer than Lowell factory women well into the 1830s. On January 1, 1837, the twenty-seven powerloom weavers who were working for Whitaker had already averaged two years employment with him and less than one-quarter had been with the company for less than one year. Among Hamilton Company weavers in Lowell in 1836, just half had been steadily employed for more than a year.[58]

Many of those who arrived to obtain work in the mills of Manayunk in the years before the Panic of 1837 had not come straight from the Philadelphia docks but had migrated from other mills in Philadelphia or neighboring counties or from textile districts in New England. Thomas and Susan Bowker and their six-year-old son went to Rockdale, in Chester County, when they arrived in Philadelphia from Manchester in 1828. Nine years later they left Rockdale, heading not for western farmlands with their pockets full of savings, as Anthony Wallace might suggest, but for Manayunk to obtain work in one of the Ripka mills that remained in operation during the depression. Israel Foster arrived with his father from Lancashire in 1823. Before he became manager of a carding department in one of Manayunk's cotton factories in the late 1830s, he made at least two moves among the mills of the area. Benjamin Gartside and his son James worked in a Blockley mill for four years after disembarking in Philadelphia from Lancashire in 1831. In 1835 they moved to Manayunk and found employment with Joseph Ripka. In the early 1840s, after years of work in the mills, John Bowker, Israel Foster, and James Gartside moved out of the wage-laboring ranks and became petty proprietors.

Many more among the mobile immigrant textile labor force, however, remained factory operatives their entire work lives. Yorkshire-born Thomas Mosely, for example, who had made his way to Manayunk in 1827 from Dedham, Massachusetts, was laboring as a spinner in Joseph Ripka's Silesia Factory a decade later. The venerable Samuel Ogden was also, as we have seen, part of the migratory early industrial laboring class. He had worked many years in English factories before immigrating to America in 1806. Over the next quarter century, Ogden, in his own words, "worked nine or ten years in the factories of the United States." In the late 1820s, he arrived in Manayunk with his aged mother and four daughters, and he ended his career there as a machinist and spinner. All of Ogden's daughters worked through the mid-1830s as powerloom weavers in Ripka's factory and no doubt supported their father who had grown too old to labor in the mills. By 1837 the daughters were on the move to other mill jobs, this time in nearby Kensington.

Other textile factory workers who lived and labored in Manayunk for only a short period had migrated from other mill towns and left for still another. John Thornily of England, for example, made his way to one of the mill towns in Delaware County after disembarking in Philadelphia in 1820. He worked in the same mill for nine years in a supervisory position. Out-migration after almost a decade of being a foreman, however, did not mean escape from the drudgery of the mills. For the next eight years, Thornily worked in at least three different mills in Philadelphia County, including nine months in a Manayunk mill. When he testified before the senate investigative committee on factory conditions in the depression year of 1837, he identified himself as a machinery repairman.[59]

The individual histories of the male members of Roxborough's new industrial laboring class varied widely, as did the amount of time they stayed in the community. They were not all "extraordinarily mobile," the term Anthony Wallace has used to describe the mill operatives of Rockdale. Nor did a high rate of mobility among many Manayunk operatives evidence a "financial ability to escape from unfavorable circumstances." Writing of Rockdale's English and Irish immigrant operatives—the core of that mill village's textile workers—Wallace has supposed that they "regarded places like Rockdale as temporary residences, where they consolidated their means and prepared themselves to move on to better jobs or even to homesteads in the west."[60] No doubt, one of the aspirations of some of the immigrants from England's industrial cities was to acquire farmland in the seemingly limitless western country. Even cheap land, however, could not be purchased on the subsistence wages of a textile operative family in the urban East. Furthermore, Philadelphia textile workers probably heard from more than one source such warnings as that directed "TO EMMIGRANTS" by the *Mechanics' Free Press* of July 5, 1828. The workingmen's newspaper advised those from England and Ireland "brought up in the Mechanical business" not to be tempted by cheap land in the West. Those "whose former habits in life neither qualify or fit them for labours in the field," the editor warned, would deceive themselves because farming in America was "arduous beyond description."

Combining what has been presented in chapter 3 concerning the general poverty and irregular employment patterns of mill workers with evidence of their transiency, it appears that Manayunk provided a starkly different industrial laboring experience than Rockdale and Lowell. Outward migration did not necessarily mean upward mobility among the early industrial laboring class. It is doubtful that Manayunk workers "used movement" to improve their situation as rural southern Massachusetts workers could and did in a labor-scarce market. A textile worker could be extraordinarily mobile and still "feel chained to the machines for a lifetime of drudgery." The experience drawn from years of labor in a number of mill towns from Lancashire to Philadelphia prompted rather than precluded class awareness and labor militancy among industrial laboring men and women. This would become evident in the events that transpired in Manayunk in the decade preceding the Panic of 1837.[61]

Industrial production had brought a new population of laborers and capitalists to Roxborough Township, sharply altering the human geography of the old flour-milling district. Manayunk's migrant, immigrant laboring class suffered material circumstances that set them apart from the traditional wage earners of Roxborough—the coopers, papermakers, carpenters, and farm laborers. The economic interests of the textile mill owners of Manayunk, in a similar fashion, contrasted with those of the employers of Roxborough—the flour millers, commercial farmers, and petty manufacturers. Once the mills were established, the industrial townspeople of Manayunk experienced poverty, propertylessness, and the concentration of economic power to a degree unknown in the prefactory township. It was, of course, not only the nature of work and production and property relations that had been transformed with Manayunk's rise. Ineluctably, the cultural and political relationships that prevailed in preindustrial Roxborough also changed.

V

The Institutions of Order

New cultural and political networks and institutions arose in Roxborough with the coming of the factory. The emergence of new churches, transportation links, and the first public school reflected the social relations and tensions of the industrializing community. Seeking political power and cultural authority commensurate with their economic position at the apex of the community, the mill owners organized churches, dominated the town council, initiated the drive for public and Sabbath schools, and guided the internal improvements of the township. Tension existed because the textile capitalists who arrived after 1820, the old elite, and laboring people held different assumptions about what the new institutions and services were meant to accomplish.

Before textiles replaced flour as the base of Roxborough's economy, the residents required few formal political or cultural institutions. Eighteenth-century Roxborough had no town council, and there were few governmental tasks that were not discharged at the county level. In neighboring Germantown, and in Philadelphia County townships in general, there was a "bare minimum of local control" and thus a minimum participation in the operation of the township in the eighteenth century. Residents of Roxborough, like those of Germantown, had little to say in the making and implementation of laws and services that affected them most: the transference of property, assessment and payment of taxes, settlement of civil disputes, and the repairs of roads and bridges.[1] Nor is there any evidence that they were interested in becoming involved in overseeing these matters at the local level.

Yet while formal political activity was absent at the community level, distinct lines of local power and authority existed. A vertical social order, manifested in associational and economic life, governed the interaction of social groups in the prefactory township. The key form of this system was paternalism, and manifested in face-to-face control, it was the essential mode of social interaction between the economic elite and the rest of the township. Not untypically, a cooper, carter, cordwainer, laborer, tenant farmer, millwright, or papermaker and his family might depend upon the same individual for a job or patronage, for housing, for access to the local church, and for poor relief.[2]

Evidence suggests that the landlord-employer/tenant-employee relationship buttressed a system of economic clientage in the community—a relationship in which economic dependency elicited social and political deference. A number of Roxborough tradesmen, particularly coopers and millers, were tenants of their employers. The wealthy cooper Lawrence Miller housed two of his fellow craftsmen in two of his three dwellings at the end of the century. Three of the large landowning farmers in the community rented out their extra dwellings to laboring men. Nicholas Rittenhouse boarded the family of papermakers who operated his mill. Tavernkeeper Nathan Levering, who owned more dwellings and acres than any resident at the turn of the century, rented four of his seven houses to unskilled workmen. Even the young laborers, papermakers, millers, and coopers who occupied the two tenement houses owned by nonresident landlords must have been aware of the powerful position of the Leverings and Rittenhouses. The names of the township's two residential enclaves—Rittenhousetown and Leverington—underscored the visibility of these families as the largest landowners and landlords in the community.[3]

The merchant millers and the Leverings—the old families who controlled a large share of the community's productive wealth—held most of the positions of influence in Roxborough's local institutions. For example, the Robesons, Livezeys, and Rittenhouses, and most conspicuously the Leverings, served as the managers of the poor and directors of Roxborough's school. More than any other family, the Leverings were involved in setting up and directing local institutions. Abraham Levering was one of the original trustees of the Roxborough public school and contributed the lot for the schoolhouse. The school remained for many years the only public building in town, and it was here, in the center of the Leverings' landholdings, that the elections for the township were held. For fifty years, from 1790 to 1840, Michael Levering was the person most frequently chosen as manager of the poor. He "devoted a great deal of his time to the poor," Horatio Gates Jones recalled, and in visiting "all who were known to be in want," Levering performed one of the legitimizing roles of the economic elite in the preindustrial social order.[4]

What did not change in Roxborough's transition from a commercial and processing economy to an industrial manufacturing one was the close connection between economic and political power. What did change, however, was the way in which authority was achieved. Face-to-face exchanges and economic clientage gave way to more formal and institutional modes of control and interaction. This process began abruptly with the arrival of Manayunk's textile capitalists and operatives.

Local government was instituted in Roxborough with the formation of the Manayunk town council in 1824. From its establishment, the canal factory owners monopolized the seats on the council. At the first town meeting, held in May 1824, cotton textile capitalists and other proprietors on the canal dominated the council offices and committees. William Brook, one of the early builders on the canal, served as the first chairman of the town council meeting and Isaac Baird, an early cotton textile manufacturer, was elected secretary. The four men who can be identified on the small ad hoc committees all owned cotton mills. The large mill owners who came to the village after 1825, such as J. J. Borie and Joseph Ripka, took an active part in town politics, as did the pioneers on the canal. Borie, who was not a resident of Roxborough, served as vice-president of Manayunk's town meeting in the beginning of 1833. Ripka's contemporaries, Charles Hagner and Horatio Gates Jones, depicted him as a vigorous participant in community affairs, describing the mill owner as "a remarkably active and enterprizing man." In Hagner's view, Ripka was "for many years the life and soul of Manayunk." Ripka led the list of Manayunk residents who petitioned for political independence from Roxborough in 1840 and was elected the borough's first burgess.[5] At least within the community that housed his profitable mills, Ripka had achieved a position of political power to match, and help secure, his economic stature.

The political interests of the industrial mill owners and the nature of social relations in the early industrial community can be seen in the drive to establish public and Sabbath schools in the township. Impetus for the movement came from the increase of laboring-class children in the township. The movement was led by the canal manufacturers, whose interest in creating local institutions that would inculcate discipline and morality in their factory labor force was not necessarily shared by laboring-class parents or by Roxborough's rural inhabitants.

Until 1836 there was no public school system in Pennsylvania. Those parents who could not afford to pay for their children's education applied to the county for aid. If the court-appointed directors declared that the parents were indigent, the child's education would be subsidized by the taxpayers. With the coming of the mills, a flood of children arrived whose education would need to be paid for by the landowners of greater Roxborough.

Charles Hagner, who moved his drug mill from the Falls of Schuylkill to the Manayunk canal in 1821, launched the effort to educate Manayunk's mill children. Hagner did not depend on the labor of children as his textile neighbors on the canal did, and he revealed a concern about the social costs of the new factory system that the more thoughtful among the antebellum capitalist class shared. Public schooling was the first among a number of causes Hagner would become active in before he moved away from "the Manchester of America" in the late 1830s. He belittled the "miserable pauper system" of education that excluded all but the most destitute families from support. He charged that the directors centered in Roxborough and Germantown were "unacquainted with the nature and necessities of the people of Manayunk" since they would not subsidize the education of the industrial laboring class's children. Because of the demand for child labor in the mills, half of the textile borough's residents were under fifteen years of age, and Hagner knew that the directors could not expect their rural constituents to shoulder the financial burden of educating the large, lower-class, school-age population of Manayunk. After successfully lobbying for representation on the county board for Manayunk, Hagner was appointed the town's first director and quickly informed the teachers and mill workers in the village that he "was ready to give orders to every applicant." In the months that followed Hagner's appointment, the Board of Control grumbled over the growing bills for Manayunk's "public scholars."[6]

Hagner claimed that his fellow proprietors on the canal were "universally ready and willing" to work together to further education in the town. Indeed, every meeting of the mill owner–dominated town council in the 1820s took up the matter of schools. And the Manayunk mill owners were responsible for establishing the first public school in the village in late 1824. Their support for public education, however, was circumscribed by their dependence upon a child labor force. Their interest in the mental and moral improvement of the children extended only to those too young to enter the factory. Asked by the senate investigative committee of 1837 if any attention was paid to children's education or morals in the mills, each witness replied negatively. Joseph Dean, who managed a mill in Blockley, added that the exception to this rule was if the children's "immorality induced them to neglect business." Joseph Ripka made clear the position of the industrial manufacturer: "It is expected that children would be educated before they are sent to work in the mills."[7]

Much as they may have wished to educate their children, the majority of laboring-class parents could not support in practice the system of public education in the township because they depended on their children's wages to supplement the family income. Some children, the principle of the Manayunk school reported, "left school for the factories as young as seven . . . and many from nine to ten years of age."[8] The breakdown of school enrollment by age in Manayunk demonstrates that few children went to school after the age of ten. Seventy-two percent of the 383 students enrolled in school in 1837 were under ten years of age. Only 43 of the pupils in Manayunk's schools were twelve or older. By 1840, four years after the public education system was instituted in the state, only 740 children attended the township's schools. They constituted just over half of all children between the ages of five and fifteen in the industrial town.

Sunday schools, like public schools, were promoted by the town council, in part because of their utility in conveying standards of behavior and morality to a young immigrant labor force. The first matter of business that the Manayunk manufacturers took up in the first town council meeting in 1824 was the organization of a Sunday school in the town. By 1837 each of the five churches in Manayunk and Roxborough operated a Sabbath school for their young members. However, these institutions of moral education, like the public schools, had only limited success in attracting working-class children. Witnesses before the 1837 factory investigation committee pointed out that less than half the children employed in the mills attended Sabbath schools in the town. Employers urged families to send their children to school on Sunday, but on the one day when they did not labor, children resisted being confined in a church schoolroom.[9]

The cultural and social networks of early American communities were reproduced nowhere more clearly than in their churches. This was true for Roxborough's preindustrial and early industrial era. The first church in Roxborough was founded by Baptists in 1789. Over thirty years elapsed before another church, established by the Methodists, took root. Then, between 1822 and 1832, four new churches were organized. We can learn much about the nature of Roxborough's changing social order through the evolution of these religious institutions.

Before 1789, Roxborough residents either held religious meetings in their homes or the schoolhouse or traveled on Sundays to Germantown or Philadelphia. Both Roxborough's Quakers and German Reformed Protestants rode up School House Lane to the meeting houses in Germantown. The Rittenhouses and Livezeys were among those who crossed the Wissahickon to join their Germantown brethren. While the Wissahickon millers worshiped outside the township, the town's Baptists gathered in the home of Nathan or Abraham Levering.[10]

In 1789, the Levering brothers organized the Ridge Baptist Church. Eleven of them headed the list of the thirty-two founding members. The family was not only instrumental in establishing the church and constructing the meeting house but, over the course of years, governed the decision-making bodies as trustees and deacons. For decades following the organization of the church, a Levering held the office of deacon.[11] The most significant aspect of the Leverings' position in church affairs, however, lay in their exercise of discipline over the communicants. The issues that occupied the monthly meetings from the beginning indicate how the authority of the Roxborough elite functioned and how it was challenged.

The major concern of the Leverings and others in the monthly meetings was the moral digressions of the ranks of the membership, most of whom were propertyless or small-propertied tradesmen or farmers.[12] The church fathers established a standing committee whose major task seemed to be to monitor the behavior of the members, both inside and outside the church. Horatio Gates Jones, whose father had been one of Roxborough's early Baptist ministers, recalled that the church was "very strict in its discipline." Indeed, the committee regularly visited and warned and occasionally excluded men and women for drinking, dancing, and failing to attend meeting. Concerned that individuals were straying from the fold, the church patriarchs adopted a policy in 1813 "to wait on" any member who missed three successive church meetings.[13]

The breadth of the Levering family authority was perhaps best understood by Baptist members like Patrick Graham and George Linn. Both worked as tanners for Nathan Levering, who owned the only tanyard in Roxborough. In 1817 the disciplinary committee charged Graham with being seen "in a state of intoxication." An employee and tenant of Nathan, Graham now faced the authority of John and Charles Levering, appointed by the disciplinary committee to look into his "present conduct of life."[14]

In spite of this display of paternalistic power and dozens of other visits, warnings, and exclusions carried out by the church patriarchs, the Leverings could only conclude that an uninterested congregation remained unimpressed by their face-to-face exhortations. In 1815, the Baptist monthly meeting lamented the "dull and lifeless state of Religion" in Roxborough and "the little attention paid to divine worship." Assuming that the problem stemmed from their nonproselytizing minister, the church leaders decided to place other preachers before the congregation in order to "promote the cause of religion amongst us."[15] Despite these measures and the addition of a Sunday school and Bible society, the Roxborough Baptist Church could barely hold on to its membership. Even after Samuel Levering was invited "to attend in the Neighborhood of the Rittenhouses for the purposes of holding prayer or social meetings" in the winter of 1818, membership lists bore out the failure to recruit new communicants. In 1820 the Baptist membership fell to the lowest point in twenty years. When the first waterpower was sold on the canal in 1821, the membership of the Roxborough Baptist Church had been declining for a decade. Although a small revival in that year brought twenty-five new members into the Baptist meeting, the congregation continued to shrink through the 1820s, as some members moved to Philadelphia and a wave of exclusions thinned the ranks in 1827 and 1828.[16]

Yet it was in the decade of the arrival of the factory to Roxborough that four new churches were established in the township. Each reflected a distinct set of cultural and social relationships in the mill town. The Roxborough Baptist Church, for thirty years the only congregation in Roxborough, would itself be transformed by the coming of the industrial order.

In 1822 the Reverend Jacob Gruber, from Saint George's Methodist Church in Philadelphia, arrived in Roxborough and organized a handful of "old-stock American Methodists" from the township and neighboring Germantown into a small Methodist class.[17] The original members, the nucleus of the Mount Zion Methodist Episcopal Church, were laboring men and their wives: William Batchelor, tender of the canal locks; William Hughes, carder; John Porter, roofer; George Jaggers, mechanic; and James Spence, storekeeper. They met in the Porters' home and then in the canal lockhouse until 1828. Consisting of 26 members, the congregation then moved its Sunday and Thursday evening meetings to the schoolhouse on Main Street, under the shadow of Borie's cotton textile factory. By 1834, the church had enough communicants to raise $950 to purchase a lot and erect a small plain church on Levering Street, one block from the canal.[18]

Laboring-class families from Manayunk's early industrial population were attracted to the Methodist theology of immediate conversion and the world of spiritual egalitarianism. We can take a snapshot of the membership in 1837, the only year for which records exist prior to 1840. Twice as many females as males appeared on the membership rolls, and the majority of those shared surnames with the males. It is impossible to establish with certainty the occupational or marital status of the women in the Methodist church. We can trace five, however, to the 1840 census, where they headed households with children and boarders who worked in the mills. The husbands and fathers of the Methodist women and children were laboring men—"manufacturers," laborers, coopers, boatmen, nailors, and masons. Only a handful made their livelihoods outside of the manual trades, and they were shop proprietors.[19]

Literary evidence also suggests that Manayunk's Methodist church was composed almost entirely of those from the laboring classes. The church "had nothing to attract to her the rich or great," Mount Zion's historian wrote, "but the earnest flow of religious fervor which marked her meetings brought the poor and uneducated sinners to her alters."[20] Thus Methodism in Manayunk shared much with its English progenitor, which, as E. P. Thompson has written, "was well suited to serve as the religion of the proletariat, whose members had not the least reason in social experience to feel themselves 'elected.' "[21] For the working-class men and women of Manayunk who made their livelihoods within an economic system that did not guarantee material security to wage laborers, the experience of shared conversion and grace must have created a powerful bond. "Which of us have not heard of the grand meetings they held upon the canal banks . . . the stirring sound of their hymns of praise?" the church's historian wrote of the Methodists of the early factory period.[22]

Recent historians have argued that in other industrializing centers such as Rochester and Philadelphia evangelical Methodism served the upwardly mobile, self-disciplined skilled craftsmen who believed grace could be continually evidenced by hard work. But for the Manayunk counterpart of this group, we must look beyond the small plain church near the canal to the Presbyterian congregation, where the ethos of self-restraint and material success underlay the ideology of the communicants.[23]

In 1832 the Young Men's Missionary Society of Philadelphia from the city's Fifth Presbyterian Church sent a layman to Manayunk to organize the five-year-old Dutch Reformed Church into the township's first Presbyterian congregation. Heading the list of members in the early 1830s were Cordelia and James Darrach, a Scots-Irish mill owner on the canal. For a year after its organization, the Presbyterian congregation worshiped in the dye house of Darrach's mill. By 1832 they had signed up seventy-nine students, many of them probably mill employees, to attend their Sabbath school. Three-quarters of the members of the church in the 1830s were females—the wives and daughters of Scots-Irish, English, and German tradesmen, many of whom had resided in the township before the arrival of the textile mills. From tax lists, we can identify the occupations of 42 percent of the forty-three male members of the church in the 1830s and the property-holding status of ten of those at the end of the decade. Five men worked in the building trades; two out of three of them were propertyless in 1840. Six of the male members worked in the textile mills; two of these "manufacturers" can be found at the end of the decade holding no property in the township. Two propertyless cordwainers joined a storekeeper, teacher, overseer, and the other tradesmen in the female-dominated congregation. The female members of the Presbyterian church likely consisted of some who worked in the mills and others, like Cordelia Darrach, from the small ownership class. They were inspired by the earnest example of the Darrachs. James, of course, had contributed part of his mill building for the church's Sabbath school, and his brother served as trustee for the Manual Labor Academy, became secretary of the Pennsylvania Tract Society Reading Room, and helped found the Flat Rock Temperance Society in 1834.[24]

The social hierarchy of the mill town was perhaps nowhere more clearly reproduced than in Saint John the Baptist Catholic Church. Jerome Keating, one-third owner of the Schuylkill Factory, organized the congregation and built the church for his fellow Irish-born countrymen and employees. Before Keating moved to Manayunk, where his father owned a great deal of real estate, the Irish laborers, mill workers, and shopkeepers made their way to Philadelphia to attend Sunday mass. Upon completion of the Schuylkill Factory in 1835, Jerome and Eulalie Keating built a mansion house at the end of the Back Row, the last string of workers' housing farthest from the mill. Here, just across Main Street, they held the first Catholic mass in the mill town. Twenty predominantly Irish-born families with a handful of German Catholics constituted the original parish. The one hundred or so early members attended services at the Keating's home until Saint John the Baptist Church was completed in 1831. The church stood on land the Keatings had donated, just beyond the final row of factory dwellings heading up toward the ridge. Within three years, the church building had to be enlarged for the rapidly growing parish of Irish mill workers, canal workmen, laborers, and shopkeepers.[25]

Jerome and Eulalia Keating, in the words of one of the early communicants, were "the soul of Catholicity" in the mill town. The recollections of the son of Francis Barat, a Keating employee and founding member of the church, illuminate the cultural authority the mill-owning family achieved within the Irish laboring-class community. Francis Barat himself instilled in his children a loyal respect for his employer. Ten years old when the cornerstone of Saint John's was laid, the Barat boy recalled Keating as "our cherished benefactory."[26] According to Barat, Keating designed "edifying customs" that legitimatized the class hierarchy of the Manayunk parish and, by extension, the social order within the mill. Until the pastor moved to Manayunk in 1833, he traveled to the mill town weekly and stayed purposefully in the "humble dwelling of one of the villagers," not "in the mansion of some wealthy person such as Mr. Keating." According to Barat, Keating and the pastor well understood the function of this act:

> They must have reasoned rightly that the poor people of the village would be all the more edified in having their pastor put up at the house of a poor hardworking villager like themselves, rather than shelter under the roof of a rich man's dwelling.[27]

The Keatings worked tirelessly within the church, in part to inculcate deference in the working-class congregation. Every Sunday between services, they each "gathered into the church pews the village children" for instruction in Christian doctrine. Barat portrayed the Keatings as a "pious, zealous, edifying couple," powerful authorities at the head of the Catholic laboring class in the community. Within six months of Jerome Keating's early death in January 1833, however, the bonds of Manayunk's Catholic community would be swiftly shattered as Irish-Catholic shopkeepers actively supported fellow parishioners in the mills in a turnout against the Catholic owners of the Schuylkill Factory.

As in the Presbyterian and the Catholic churches, the industrial capitalists of Manayunk played a visible role in the formation and direction of Saint David's Episcopal Church, located at the northern end of the village away from the mills on the canal. The few extant membership records of Saint David's suggest that many Protestant mill owners, their employees from Lancashire and Yorkshire, and local skilled tradesmen worshiped in this church. At the end of 1831, the Reverend Robert Davis, who had organized the First Episcopal Church in Philadelphia, contacted Charles Hagner to help him establish a congregation in Manayunk. He canvassed the community and located, he claimed, 300 persons who had been brought up in the Episcopal faith. Within a week of Davis's inquiry, he and Hagner had organized a parish that included Joseph Ripka, the mill-owning Wagner brothers, and Hagner himself as founding vestrymen. Joseph Ripka donated a lot not far from Rosenberg for the church, and his fellow textile manufacturer Tobias Wagner funded its construction. Those who filled the pews were probably, like the founding members, manufacturers, farmers, and small proprietors. Those who served as vestrymen were storekeeper Joseph Clark, teacher and magistrate Francis Murphy, machinemaker Joseph Haywood, Ripka's clerk and confidante John Koch, and four of the factory owners on the canal.[28] The Episcopal church, more so than any other congregation, served as an institution that integrated Roxborough's established tradesmen and proprietors and Manayunk's textile capitalists and loyal employees.

High up the ridge away from the mill town on the Schuylkill, the four-decade-old Baptist church remained a congregation of old Roxborough laboring and craft families until the early 1830s. None of Manayunk's industrial mill owners appeared on the membership lists of the church. The landholding craftsmen of Roxborough—the Conrads, Holgates, and most prominantly, the Leverings—continued to serve in the leadership positions of the Baptist meeting as their fathers had done since the late eighteenth century.

As new churches were established in Manayunk to meet the demands of a diverse and expanding population, Roxborough Baptist Church stagnated. In 1828 a new minister, Samuel Smith, baptized fourteen new members, bringing in the first group of communicants from Manayunk. Horatio Gates Jones recalled that he began to see many strangers at the church. In the winter of 1830–31, however, following the destruction of the meeting house by fire, the congregation was only ten members larger than in 1821. But in the fall of 1831, the fate of the Roxborough Baptist Church turned with the hiring of a new pastor, Dyer A. Nichols, who began a series of evangelical revivals that peaked in the winter of 1832–33. With the addition of over 200 new members in those months, the old church on the ridge suddenly became the largest congregation in the township.[29]

The dynamics of the Baptist revival relate to a broader social process in the mill town, which will be considered in following chapters. It is clear, however, that the old church's abrupt expansion after forty years, as well as the formation of the other congregations, was connected closely to the forging of the factory-based society. With the establishment of the mechanized factories, the old and new residents of Manayunk formed networks heretofore absent in the community. Within the associational institution of the church, tradesmen, laborers, factory workers, petty manufacturers, flour millers, and textile mill owners carried out the search for cultural autonomy and authority in their transitional society.

One other aspect of community life—the politics of internal improvements—allows us to chart changes in Roxborough's early industrial social order, especially as it involved a contest between the old merchant millers and the new textile capitalists. The location of roads between mill and market or tavern and turnpike could determine the economic fate of individual craftsmen, tradesmen, and manufacturers. Formal appeals to the county for the construction of roads or bridges thus easily became controversial issues among residents of Philadelphia's surrounding townships. The records of court road petitions filed by Roxborough residents provide an unusual view of competing interests and relationships of local power in the expanding market economy. From the eighteenth century into the factory period, the powerful commercial and manufacturing capitalists were continuously involved in the politics of road building.

Through the eighteenth century, the merchant millers of the Wissahickon obtained public roads to connect their valuable mill seats with the main turnpikes into Philadelphia. Toward the end of the century, however, the Rittenhouses and other Wissahickon millers began to meet with growing opposition from Roxborough inhabitants. At the head of this opposition were the Leverings, whose interest lay in developing the western section of the township. Provincial localism explains in part the enmity toward the Rittenhouses expressed in the road petitions of the latter part of the eighteenth century, for the Rittenhouse family belonged more to Germantown, where many of the family members lived, than to Roxborough Township. Their lands lay along the border of the two townships. They intermarried with the Pauls, Deweeses, and Gorgases, all Germantown mill-owning families and "the most aristocratic members of their community." The Roxborough Rittenhouses attended their Germantown brothers' churches and sent their sons to the Germantown school.[30]

The Rittenhouses had initially established the commercial links between the Wissahickon mills and Germantown. Before the Revolution, a court order noted, "a very frequented Mill Road [led] from the Germantown Main Street to [Henry Rittenhouse's] Grist Mill on Wissahickon Creek."[31] In 1791, however, an appeal for another mill road drew to the surface the communal ire of Roxborough. In December of that year, Martin Rittenhouse filed a petition with the court arguing that "your Petitioners Labour under great Inconvenience for the want of a Convenient Road to Accommodate the said Mill and the Inhabitants of Roxborough . . . To Travel pass and Repass to and from Germantown to the Ridge Road."[32] The proposed road would cut through the lands of John Levering and Michael Righter, who together controlled a wide tract of land running from the west side of the Wissahickon to the Schuylkill.

Within a few months the landholders and tenants of Roxborough, "a very few excepted," filed a "Memorial and Remonstrance" against Martin Rittenhouse and the handful of endorsers. The counterpetition expressed the hostility of the community toward the merchant millers on the Wissahickon and articulated the economic interests that were at issue:

In the space of less than four miles there are five public roads;

which are supported by us at a heavy expence, for the benefit of the Merchant Millers on the Wissahickon [;] all of them communicate to Germantown.[33]

The petitioners noted how one of these roads had "been constantly used for many years to pass and repass to and from Martin Rittenhouse's Mill." They protested paying the "heavy additional road tax," principally for the benefit of the millers. Speaking to the interest of the community, the petitioners argued that the "too thin" woodlands would be further cut and the "too small" cleared parcels again subdivided when the road was constructed.

The petition characterized the merchant millers as an unscrupulous elite trying to benefit themselves at the expense of their community. The Rittenhouses had indeed used economic coercion and family connections to obtain support for their project. The counterpetitioners charged that Martin Rittenhouse had gone outside of the township to procure signatures: of the twenty-eight signers of the Rittenhouse petition, less than half resided in Roxborough, "and a moiety at least of those are of the same family . . . Such an attempt to deceive, by mistaking facts in the first place, and then borrowing names to confirm the deception," earned the Rittenhouse family the contempt of their township.[34]

The alignments on the opposing petitions reflected the pattern of economic clientage in the prefactory community. Rittenhouses from Roxborough, but mostly those from Germantown, signed Martin's petition. Of the handful of inscribers who were not either Rittenhouses or from Germantown, two can be identified as employees of one of the Rittenhouses. Michael Ash, young and single, lived and worked as a laborer for Abraham, owner of a flour mill. Jacob Markle, a papermaker who also signed the petition, was Jacob Rittenhouse's tenant and managed his papermill. The counterpetitioners also shared economic and residential links. Heading the list of complaintants were the Roxborough landholders and their tenants through whose land the road would run. It is likely that the propertyless and small propertied craftsmen and laborers inscribed their names under the Leverings and Righters of the western side of the township because of their hostility to the merchant millers and because they clearly understood the nature of their dependency on their employers and landlords.[35]

A battle over the establishment of a new road in 1811 again pitted the Wissahickon millers against the community. In June 1811, landowners of the northwest section of Roxborough petitioned for "a road or cart way" from John Wise's mill on the Wissahickon to the Ridge Road. The majority of the petitioners were among the township's largest property holders, including Nathan Levering and members of the farming elite. They were joined by a handful of substantially propertied craftsmen-manufacturers. The route they endorsed would shorten and improve access from the Levering-dominated section of the township to the Germantown Pike.[36]

At the following Court of Quarter Sessions, merchant millers John and Joseph Livezey filed an opposing petition for review. The Livezeys, who had built a mill road connecting them to the Ridge Road and the Germantown Pike, argued that three public roads leading to and from John Wise's mill to the main roads already existed. The proposed route would front the stream on the Livezey property, and, the millers complained, would "deprive [them] from improving a valuable waterfall for erecting Factories of any kind."[37] Their appeal was met with a petition for re-review, which contained twice the number of signatures as the initial petition. This time the inscribers justified their request in terms of the public good, what was "most convenient to the Neighborhood in general and of the greatest Public Utility."[38] For the next decade, the Livezeys unsuccessfully contested the John Wise mill road and its improvement. At no time did fellow Roxborough residents endorse their memorials.

As in the dispute over the Martin Rittenhouse mill road two decades before, the inhabitants of Roxborough aligned with the Leverings against the merchant miller. Both controversies over transportation links exemplified the nature of authority in the community. The Livezeys, like the Rittenhouses, had strong ties outside of their own township, particularly in Germantown. The Livezeys belonged to the Germantown Quaker meeting. They owned property in that township. One of their large flour mills lay on the boundary line of Roxborough and Germantown. In the 1811 dispute, when the Livezey brothers attempted to protect their mill property, they were unable to garner the support of any of their fellow townsmen.

The placement of roads and bridges in the township charts the shift of the center of development away from the mills on the Wissahickon to the banks of the Schuylkill even before the completion of the canal along the river in 1819. In the eighteenth century, the township built most of its major roads between the Wissahickon mills and Germantown. School House Lane, for example, the second public road in Roxborough, ran from the Market House in Germantown to Robeson's mill at the mouth of the Wissahickon. By the early nineteenth century, the road petitions indicate that Roxborough inhabitants concentrated on improving communication links with Philadelphia and towns along the Schuylkill. The first bridge to connect Roxborough with the western side of the Schuylkill was completed in 1810. In the same year, nearly one hundred residents of Philadelphia and Montgomery counties filed a petition stating they were "under the great inconvenience for want of a Publick Road" along the east bank of the Schuylkill in Roxborough.[39] The completion of that road, the Ridge Turnpike, in 1812 ended the days when residents of the township had to go to Germantown to the pike in order to go to Philadelphia. Not surprisingly, Nathan Levering, whose lands and businesses bordered the ridge route, had been instrumental in organizing the improvement of the turnpike to Philadelphia.

The construction of a section of the Schuylkill Navigation Company Canal on the township's western border in 1815 spelled the eclipse of the Wissahickon millers' economic domination. As we have seen, the mill town of Manayunk grew at a phenomenal rate after the completion of the Flat Rock Canal and mill seats in 1820. In this period of vigorous development, the politics of road building continued to reflect how control of property and production was translated into political power. In the early industrial era, it was the cotton textile capitalists of Manayunk who financed and fostered transportation links that would improve access between their markets and mills. Indeed, few roads or bridges were developed in Roxborough after 1819 that were not in the mill borough or on the west side of the township. Thus the Wissahickon section of the township became further isolated from the new industrial order.[40]

In 1825 the most direct route between Philadelphia and the manufacturers on the canal was the Manayunk Road, which often became impassable during the wet seasons. To better the route between the factories and the center of the port city, the mill owners obtained an act of incorporation to build a macadamized toll road. "No one thought it would ever pay any interest, but it was necessary for our business," explained Charles Hagner.[41] Five of his fellow mill owners served as commissioners on the seven-member board to raise subscriptions and direct the local improvement. Believing that the mill owners should contribute to the development of Manayunk's main route, the commissioners assessed each of their fellow industrialists a share of the twenty dollar subscriptions.

The pattern of alignments over the public improvements in the 1820s and 1830s was reminiscent of those at the turn of the century. In 1828, when the canal manufacturers led about ninety taxable men in Manayunk in a successful petition for construction of a bridge over the canal and a road joining the mills and Schuylkill Road, the Wissahickon merchant millers were conspicuously missing from the list of inscribers.[42] And when the Wissahickon milling families sought approval for a public bridge at Enoch Rittenhouse's mill between Germantown and the Ridge Road in 1828, the court concurred that it was too expensive and denied the petition. Indeed, throughout the early 1830s all but one of the new public roads or improvements branching from the Ridge Road ran west toward the canal.[43]

The most important transportation link that Manayunk gained at the expense of the Wissahickon inhabitants was the route of the Philadelphia, Germantown, and Norristown Railroad. The location of the track remained an open matter when plans were announced in late 1830 to begin construction of the Philadelphia to Roxborough link. The *Germantown Telegraph* of November 10, 1830, noted that the possibility of a railroad route along the Wissahickon greatly excited those inhabitants "whose lands will be enriched thereby." Those who had the most to gain, the editor pointed out, were the "numerous millers along the creek, who [would] secure . . . a readier and cheaper conveyance than formerly for their flour to the Philadelphia market." The railroad's board of directors, however, decided to lay the route through Manayunk, close to the canal. The Wissahickon millers, disappointed in losing the bid for the railroad in their vicinity, did not likely join in the celebration of the opening of the Manayunk depot, two blocks from the textile mills, on an October day in 1834. Among those who did no doubt partake in the festivities were the new industrial manufacturers—"the capitalists," as Horatio Jones described them, "who felt that Manayunk was now close to the metropolis."[44]

As workmen laid the tracks along the canal toward the mill town in 1833, the Wissahickon millers appealed to the courts for approval of a public road to wind the length of the Wissahickon Creek from the Rittenhouse mill road to Germantown—a feeble substitute for the modern transportation link that the town of Manayunk had obtained. Even this was denied them, however, as hundreds of petitioners from the area countered their memorial. Mirroring the contempt for the merchant millers expressed in the Rittenhouse road case four decades before, the 1833 counterpetition decried an isolated elite who would use public monies to advance private gains. Roxborough's citizens claimed that they were already burdened by taxes "in consequence of the great number of roads which they have to keep in repair for the accommodation of the Millers along the Wissahickon." They chided those of narrow self-interest who would take advantage of the community—the "not more than five or six individuals" who already had adequate roads "to convey their flour to market."[45] The pattern of opposition and hostility toward development outside of the manufacturing center, amid steady road, bridge, and railroad building in Manayunk, contributed to the disaffection of Roxborough's old elite from the industrial capitalists in their township. It was such disaffection that would drive them to political organization in the 1830s.

VI

PRIMITIVE PROTEST, REPUBLICAN REFORM, AND RELIGIOUS REVIVAL

Beginning in the late 1820s, work, politics, and religion became arenas of social conflict in the community of Roxborough and its factory center of Manayunk. At the core of these struggles was a questioning of the industrial system of production and its impact upon the lives of farmers, flour millers, tradesmen, laborers, millworkers, and handloom weavers. The cotton factory—"the physical instruments of production"—symbolized for these people the forces that had given rise to new social relationships and conditions in their community. Consequently, political and religious tensions, as well as labor-capital conflict, revolved about the mills. To paraphrase E. P. Thompson, the community of Manayunk entered a crucible in the 1820s and emerged, after a decade of popular agitation, in a different form.[1] The remaining chapters will trace this social and political process that accompanied early industrialization.

The initial challenge to early industrial capitalism took the form of violent assaults by Philadelphia handloom weavers on the mills themselves. Members of one of the city's oldest and largest textile trades, the weavers played a prominent role in the popular protest connected with Manayunk's industrialization. Their opposition took the form of a series of machine-breaking incidents and mill burning, which began soon after the first powerloom factories were constructed on the Manayunk canal in 1821. In 1823, fire destroyed two mills in Manayunk containing spinning machinery and powerlooms. "There is every reason to believe it is the work of an incendiary," a correspondent to the *Democratic Press* wrote on March 7. Machinery "of the most improved kind," worth nine to ten thousand dollars, went up in flames. In the 1830s, John Winpenny's mills burned to the ground three times. "There can be no doubt," the *Germantown Telegraph* reported on the blaze that destroyed his mill in 1839, "that this fire was the work of an incendiary."[2]

By 1830, assaults on the factories of Manayunk had moved beyond surreptitious acts of arson. In the same manner that the night raids of the blackened-face General Ludd gave way to Captain Swing's daylight marches, bands of weavers with clubs, guns, kindling, and torches openly marched on Manayunk to destroy the powerlooms and mills that housed them. The first recorded incident of such overt opposition occurred in 1830 when James Kempton introduced Alfred Jenks's new powerlooms, which were capable of weaving colored checks. Handloom weavers and "others opposed to labor-saving machinery," Philadelphian Edwin T. Freedley recalled in 1857, went to the Kempton Mill "with the avowed purpose of destroying it." They were stopped, according to Freedley, by an "armed force of men." Six years later a similar course of events unfolded when Joseph Ripka built his third mill and installed 600 thirty-inch-wide powerlooms. A "large mob" of Kensington handloom weavers headed for the mill but were turned back by "a military company from Manayunk."[3] As the depression of 1837–42 ground to an end, Kensington weavers again found cause to attack the mills of James Kempton, who began manufacturing a cotton article "by a much cheaper process" than the handloom. On the night of September 23, 1842, Manayunk's sheriff had to rouse Kempton to arm himself and his friends in defense of his factory against an estimated twenty-five Kensington handloom weavers. At 1:30 A.M., while Kempton's allies patrolled the roads, seven constables encountered a group of fifty to sixty Kensington Luddites. Lookouts began to ring the factory bells, causing the band to return to the city, but not before two constables had been shot. Although this latter incident occurred outside the time frame of this study, it attests to the endurance of a traditional strand of popular protest among Philadelphia's handworkers.[4]

Certain factors confirm that Philadelphia's handloom weavers, whose livelihoods were threatened by the powerlooms, were responsible for both surreptitiously burning the mills and marching as a crowd to destroy them. First, the marginality and erratic state of their occupation created a fierce collective consciousness among weavers as wage earners, and often as members of an immigrant-dominated craft.[5] In the summer of 1828, the *Democratic Press* briefly described a "large riot" of Irish-born weavers in the Northern Liberties that left one watchman dead. The street fighters made known their mutual interest when they seized a dwelling in the embattled section of the city and proudly raised the flag of the weaver's trade. In a separate incident, ethnic conflict provided another display of camaraderie among the immigrant weavers. When nativist foes again attacked Irish weavers who had congregated in a Kensington tavern, they occupied the second story and unfurled the weaver's banner in the midst of the assault.[6]

Second, the Anglo-Irish handloom weaver carried to early industrial Philadelphia a tradition of machine breaking and Luddism. In 1811 and 1812, Lancashire and Cheshire handloom weavers and other tradesmen burned the warehouses and mills of the first manufacturers to use powerlooms in those textile districts. As the machines spread rapidly around Manchester in the mid-1820s, handloom weavers again took direct collective action. In the depths of the depression of 1826, Blackburn weavers began a movement of machine breaking that left "not a single powerloom" within six miles of that town before spreading through East Lancashire and into Manchester.[7]

If the ideology of General Ludd had been transported across the Atlantic, its reemergence as a strategy of protest in Philadelphia textile districts was probably influenced by an indigenous critique of labor-saving machinery that implicitly sanctioned machine destruction. Such an ideology was popularized by William Heighton, the radical English-born cordwainer who founded the Mechanics' Union of Trade Associations (MUTA), and who in 1828 began publishing the *Mechanics' Free Press*. In the editorial pages of his paper, Heighton articulated for Philadelphia's working people a critique of capitalist society that included an indictment of labor-saving machines. Implicit in Heighton's editorials and pamphlets was a rationale for Luddism. Heighton railed against those (particularly legislators) who encouraged the invention of labor-saving devices "under the influence of commercial competition" rather than out of a concern to limit the hardship and hours of labor. He called for penalties against the inventors of machines that displaced workingmen.[8] In the *Mechanics' Free Press* of September 27, 1828, drawing upon the labor theory of value, Heighton published "The Evil Effects of Labour-Saving Machinery Remediable," an editorial that spoke to the immediate plight of the Philadelphia handloom weavers vis-à-vis the mechanized powerloom:

> If, by the introduction of Labour saving Machinery, a number of Operatives are thrown out of employment, or in any material degree injured in their trade, such operatives have as just a claim on the profits which were the result of the improvement or remuneration for losses sustained thereby as does the landholder for his lands, which are taken for construction of a canal.

There is no direct evidence that urban weavers, possessing a history of militant direct action, saw such editorials. But in view of the fact that open assaults, as opposed to anonymous acts of arson, followed in the wake of publication of Heighton's paper, we can speculate that weavers were inspired by the editorials.

Like their English counterparts, Philadelphia's handloom weavers were loath to join the organized agitation of factory textile operatives to prevent wage reductions. Such efforts were irrelevant "to a trade already in the lowest possible state of degradation, as a result of the introduction of powerlooms."[9] Machine breaking proved the most violent form of working-class opposition against the early factory system in Philadelphia, and the most relentless. While handloom weavers never entered the Manayunk mills as wage earners, they set forth a distinct and persistent protest against what they perceived to be the injustices of industrial production.

A different form of conflict between labor and capital took place inside the mill gates in the 1830s as the first generation of industrial workers challenged the prerogatives of capital to set wages, hours, and working conditions. To begin to explain both the ideology and the opposition of Manayunk's factory operatives—the mule spinners and powerloom operators—that emerged in the late 1820s, it is necessary to describe the particular dilemma of the early industrial capitalist. As in England, cycles of expansion and depression characterized the Mid-Atlantic cotton textile industry in the third and fourth decades of the century. The highly competitive character of the industry led to bursts of investment and increased production, then to falling prices and suspended investments. In the 1830s, while prices for finished goods fell, textile capitalists like Borie and Ripka could chart the steady rise in the cost of raw cotton. Hence periodic contractions plagued the Mid-Atlantic textile industry, plunging textile workers and others dependent upon this manufacturing sector into months of unemployment almost every two years.[10]

Newspaper accounts and reports filed by benevolent institutions chronicled this cyclical pattern in the cotton industry during the decade preceding the Panic of 1837. The *Niles' Weekly Register* of May 26, 1832, reported that "many of the cotton and woolen manufacturers have in part, or altogether, suspended operations in the neighborhood of Philadelphia . . . ; the want of employment will be hard to bear." On February 5, 1834, the *Germantown Telegraph* revealed that some factory proprietors relinquished cash payments for sixty-day due bills to ride out a recession and prevent the closing of their businesses. Unemployment predominated again in the textile industry in 1835. The Union Benevolent Society in the city reported that because of the "curtailment of business and want of employment," the number of their charges had greatly increased during the winter of 1835–36.[11] Each recurring recession hit hardest those industrial centers like Manayunk that contained numbers of tradesmen and proprietors who were dependent upon the selling of commodities and services to a factory-employed population.

The principal tactic of the cotton manufacturer to retain a profit margin in the face of falling prices was to reduce the wages of his labor force. It was the periodic reductions of wage rates made by the mill owners that triggered the mill strikes in Manayunk. The first occurred in the fall of 1828 and was led by the mule spinners of the Schuylkill Factory, owned by Jerome Keating and J. J. Borie. They joined fellow cotton spinners throughout the city and in Norristown in the first general turnout by mill operatives and the first strike by any trade against wage reductions in the region.[12]

They did so without the support or guidance of William Heighton's union. Formed in 1828 and described as "the nation's first bona fide labor movement," this association of skilled craftsmen excluded factory and other types of laborers, concentrating instead on improving the economic situation of journeymen in the large urban trades.[13] While the cotton spinners and less skilled mill operatives of Manayunk and other textile districts did not belong to the MUTA, it would be mistaken to assume that they were bereft of craft cohesion and trade organization.[14] When the strike of 1828 by Philadelphia and Manayunk mule spinners interrupted two years of industrial peace in the city, Heighton and fellow labor leaders, who had worked assiduously to preserve that peace, could no longer ignore the oppositional force of the factory spinners.

Heighton underestimated the tradition of trade union radicalism among the textile factory spinners because he ignored the transatlantic history and experience of these workmen. In Lancashire, where the takeoff of capitalist industrial production preceded America's by about a quarter of a century, and from which many of Philadelphia's mill workers emigrated, mule spinners had been organizing and conducting strikes since the 1790s. When Manchester cotton spinners made their first concerted effort to oppose wage reductions in 1818, they possessed a history of active trade unionism going back twenty-five years. In 1826 they sustained a year-long series of strikes against wage cuts. Thus, when Borie's immigrant mule spinners turned out against his proposed reductions, they acted within a thirty-year tradition of factory labor opposition against threats to their livelihoods.[15]

In September 1828, the Manayunk cotton mill owners announced a 25 percent reduction in wages for all operatives, blaming the current price depression in finished goods. The highly skilled mule spinners resisted. They proclaimed that, notwithstanding the recession, they each "earned for the mill owners, from $40 to $50 per week." There was no legitimate reason, the spinners believed, that their wages should be cut, given what the "avarice" of employers had "already accumulate[d] in the form of profits." Manayunk's spinners formed the Journeymen Cotton Spinners Society and obtained support from the cordwainers' and carpenters' trade societies in Philadelphia, who raised funds to support their standout "against the oppressive encroachments of their employers."[16]

The financial support of some of the city's organized trades that sustained the strike was countered by Borie's determination to hire strike breakers and lock out the militant mule spinners. In November, the striking mule spinners turned to force to protect their jobs, attacking or threatening to harm those seeking employment at the mill. On November 7, five of Borie's former mule spinners met Edward Kennedy as he was leaving the Schuylkill Factory. They admonished him that he should not "make a bargain" with Borie: "There were men enough there, and . . . it was hard to be taking the bread out of other men's mouths." The strikers warned Kennedy of their determination: "As sure as he should engage himself as a spinner," spinners John Marshall and J. J. Treillou threatened, "so sure would he lose his life." They would have "their own wheels back," they announced, or they would "lose their lives in the attempt."[17] Material concerns certainly underlay the Manayunk mule spinners' fierce commitment to their factory jobs. Their actions were also sanctioned by customs of their trade that disparaged strike breaking. In 1792 the Manchester mule spinners had codified a prohibition against strike breakers when they formed the first regular organization among members of their trade. They adopted a written rule that proscribed members from working in a shop where a strike had taken place.[18]

To combat the direct action of the militant mule spinners, Manayunk's mill owners drew on their own body of tradition, embodied in the English common law of criminal conspiracy. It had been applied in the Cordwainers Conspiracy cases between 1806 and 1815 when American employers first resorted to the courts to protect their interest against organized actions of their employees. In 1818, 1821, and 1823, English authorities had used the Combination Laws and the common law definition of conspiracy to break spinners' strikes in Stockport, Blackburn, and Bolton.[19] Likewise, Borie used the courts to destroy the mule spinners' organized stand against him. Their strike apparently was dealt a fatal blow by the decision of the Philadelphia County Court to bind over Marshall, Treillou, and a fellow spinner because of the incident at the gates of Borie's mill. "All parties concerned ought to be convinced," the presiding judge declared, "that combinations and conspiracies of this character are illegal." The judge warned in his ruling that such unions of labor were dangerous: "In our country, but more especially abroad, combinations like these have led to consequences the most disastrous."[20] The state's conviction of the militant spinners for conspiracy helped to end the three-month-old impasse. The spinners returned to work in December after accepting a 10 percent reduction in wages.

With the state reaffirming its willingness to intervene on the side of employers, the Manayunk mule spinners' strike in the fall of 1828 galvanized the town's mill owners. William Sullivan has noted that the spinners' strike of 1828 "marked the beginning of the aggressions by the employers," who, over the next two years, imposed successive reductions totaling 30 percent.[21]

In Manayunk, this aggressiveness took the form of alliances among the textile capitalists. Borie himself led the campaign for cooperation among Manayunk mill owners to keep wage rates down. In December 1829, learning that a fellow manufacturer intended to raise the wages of his powerloom weavers to the earlier higher rates, Borie spelled out the mill owners' need for a united commitment to the successive wage cuts that began in the previous year. The canal mill owners were apparently having difficulty procuring women to operate their looms, and one of them, Mark Richards, argued it was necessary to increase wages to vie for the scarce hands. Reminding Richards that in order to compete, the other mill owners would have to match the increase, Borie urged him to forego his decision. The workers themselves believed that business had improved enough to warrant raises, and Borie cautioned that once Richards raised his weavers' rates, the mule spinners and the other operatives would also request increases. To restore wages to the former rate would also be a strategic error, for the mill owners, having exhibited weakness, would confront greater opposition when they next wanted to affect a wage reduction.[22]

Thus Borie articulated one of the vexing imperatives of the competitive cotton industry: uniformly higher wages in Manayunk would have "no bearing on more distant factories whose wages will continue on the reduced scale." Paying lower wages, manufacturers in other areas would undersell their Philadelphia counterparts in the cloth market. "We shall be less able to compete with them," Borie argued, "and competition amongst ourselves is nevertheless the great cause of our bad business." In understanding their need to cooperate as employers and manufacturers, and in acting in unity to preserve their profit margins, the early industrial capitalists of Manayunk were realizing their common interest. To remain competitive with mill owners in other textile districts, they had to learn to stifle competition within their own.[23]

The Schuylkill Factory strike of 1828 also provoked an awareness among Roxborough's wage earners and petty proprietors of a common bond they shared with the mill operatives. During the conspiracy trial of the striking spinners, eight Manayunk mule spinners were joined by five Manayunk tradesmen in testifying in the defense of Marshall and Treillou. They backed up the spinners' alibi, rejected by the court, that they had been in Philadelphia, not Manayunk, at the time of the purported threats of violence.[24] In the months that followed this limited expression of working-class solidarity, a growing number of men in the traditional trades, along with a few shopkeepers, began to question the incongruities of industrial society and to reproach the power of the industrial capitalists. They formed themselves into a local branch of the Workingmen's Party and looked to legislation to correct the abuses.

Philadelphia's Workingmen's Party was an offshoot of William Heighton's Mechanics' Union of Trade Associations. In an address to Philadelphia's crafts in November 1827, Heighton summed up the nature of the dilemma facing the working class. While making an appeal for support of his trade union, he explained that the workingmen's problems—and salvation—lay within the structure of democratic politics: "All our legislators and rulers are nominated by the accumulating classes and controlled by their opinions—how then can we expect that laws will be framed which will favour our interest?" The answer was to purify the perverted republican system and form a party that put workingmen in the seats of power now occupied by men hostile to their interests:

> Then let us resolve to never more give our suffrages to any but members of the working class, or to such others as will publicly pledge themselves to support our interest in the legislative hall.[25]

In July 1828, Heighton's union gave birth to the Workingmen's Party of Philadelphia. It advocated, unlike the Jacksonian and National Republican parties, a range of economic and social reforms. The short history of this party (which ran its last slate in 1831), the occupational base of its constituency, and its relationship to the Jacksonian and anti-Jacksonian parties have been well laid out, though along different interpretive lines. What will be explored in the following pages is why the issues and ideology of the Workingmen's political movement proved receptive, not to factory operatives, but to the tradesmen and shopkeepers of Roxborough, who developed one strand of opposition to industrial capitalism.[26] Ironically, it was Manayunk factory operatives, not the town's tradesmen, whom Heighton initially hoped to bring into the party in the summer of 1828. In the *Mechanics' Free Press*, he made a direct appeal to "the Operatives of the City and County of Philadelphia," those at the bottom of the order among the producing class who were "commonly designated *lower class, swinish multitude, Gibeonites, Canile,* &c." "Let us convince our oppressors we will no longer suffer ourselves to be made the tools to elevate them by our votes," he exhorted, while arguing that the Bank of the United States depressed the value of their wages by circulating paper money.[27]

Textile operatives, however, were undoubtedly concerned with the issue at hand not explicitly addressed by Heighton: wage reductions. In August, when Heighton formally called for the county's mechanics to assemble their delegates to represent the working classes, it was not the factory operatives of Manayunk who organized but the mill town's skilled tradesmen and petty proprietors. James Renshaw, an innkeeper in Manayunk, hosted the first meeting of the Workingmen of Manayunk, and Leonard Allen, a machinemaker, and William T. Young, a shoemaker, served as chairman and secretary of the organization. No spinners, manufacturers, dressers—those from the ranks of operatives—or laborers tarnished the list of self-described gentlemen at the meeting. Men from the building crafts—carpenters, a mason, a turner—joined two blacksmiths and machinemakers on the committee that would actively campaign in their community for the Workingmen's ticket.[28]

Undaunted by a poor showing of their candidates in the election of 1828, which swept Jackson's local slate into power in the county and city offices, the Workingmen of Manayunk joined forces with the Workingmen of Roxborough the following year to support a slate in the assembly elections. The ranks of the Workingmen's Party of the township had tripled by the fall of 1829 when they met in the home of one of the Leverings to nominate delegates for the county convention. Skilled tradesmen again dominated the party organization. Typically, they or their fathers had carried on a livelihood in Roxborough before the building of the textile mills. Those that can be traced to the 1822 tax list owned property in the township. Many had worshiped together at the long-established Baptist church, including Benjamin Miles, who built the first blacksmith's shop in Manayunk. There is no evidence that the textile factory operatives or unskilled laborers of the township were part of this expanded "workingmen's" organization.[29]

The issues supported by the Workingmen of Roxborough and the rhetoric of their resolutions reveal the direct line these skilled tradesmen drew to the ideals of 1776. Implicit in their ideology was a rejection of the cultural and political world of the mill town's growing industrial proletariat. They expressed these ideals when they read an address penned by their counterparts from the Northern Liberties at a meeting in the fall of 1829. A large crowd at Levering's home heard that "the time has arrived when the Working people are determined to assert and maintain their natural and political rights." Evidence that those rights were being abrogated lay in the numerous "perverted" laws and legislation "passed for the benefit of the rich" and against the interest of the working classes. These included the evil banking system, the incorporation of monopolizing companies, and the failure to pass a law for public education. Redemption, as any true republican knew, lay within the political system, of which the working men had not taken advantage. "To the Elective Franchise, then, must we look as the redeeming power that will improve our condition." These mechanics did not propose a redistribution of economic power, but only that workingmen share in the decisions that affected their material and social well being. They objected to a rapidly growing inequality of wealth—a situation that was not consistent with the republican worldview. It was in "the important duties of the ELECTOR," they stated, "that we can stand on a proud equality with wealth and aristocracy."[30] To promote what they perceived to be the common causes of the working people—abolition of licensed monopolies, public education, abolishment of imprisonment for debt, and a mechanic's lien law—the Workingmen of Roxborough Township consciously adopted the tactics of their patriot forebearers by appointing a committee of correspondence and a committee of vigilance.[31]

In the election of 1829, the Workingmen's candidates received fewer votes in Roxborough than in the previous year. Jacksonian candidates attained a slim majority over the anti-Jacksonians (National Republicans) in the county contests. Undaunted, the Workingmen of Roxborough organized again for the election of 1830, but "Workeyism" and "Federalism," the Jacksonians triumphantly proclaimed, went down to another defeat that year. At this point the old stock tradesmen of Roxborough began to move into a formal alliance with the National Republicans, the "forces of Federalism."[32]

There is no evidence before the presidential election year of 1832 that a formal Jacksonian or National Republican party organization existed locally in Manayunk or greater Roxborough. We can begin to understand the composition of the Jacksonian voters in the mill town, however, as early as 1828. Early in that year, a member of the Workingmen's Party of Roxborough, schoolteacher Francis Murphy, wrote to a friend in Ohio about a "Little Politicks" in the town: "every drunken fellow we meet here [stumps] for Jackson and it is doubtful whether he will not be elected President." In a 4th of July oration of that same year, John Elkington, also a member of the Workingmen's Party of Roxborough, announced that "some have it, that a town called

Jacksonville, is in agitation, at the upper end of Manayunk."[33]

The pro-Jackson, and at times intemperate, laboring men of Manayunk no doubt participated in an Independence Day celebration a year later, and their rituals provide a glimpse of how widely the cultural and ideological worlds of the Roxborough Workingmen and the Manayunk laboring men diverged. The day's events were recounted in the *Democratic Press* of July 15, 1829. First, the Manayunk Rifle Company marched to a hotel in the mill town to hear one of their members read the Declaration of Independence. An evening of food and drink followed with at least one hundred militia men and other residents sitting down to a dinner hosted by "Mr. H." (undoubtedly the fervent Protestant "Friend of Ireland" and Jacksonian, Charles V. Hagner). It is doubtful that the thirteenth toast would have been presented at a meeting of Roxborough's Workingmen, of which few Irishmen were members: "Emigrants from all climes; may their most sanguine expectations be realized on reaching Columbia's shores—*Erin go bragh.*"

Judging by the programs of the two parties, it makes sense that Manayunk's Catholic and Protestant factory laborers would be drawn to the Jacksonian and not the Workingmen's ticket. For Manayunk's operatives and day laborers, many of whom had emigrated from Ireland, and many of whose children needed to work in the mills to support the family economy, the humanitarian and social reforms of the Workingmen's Party platform likely seemed irrelevant if not inimical to their cultural and economic interests. As Arthur M. Schlesinger, Jr., has put it, "the main workingmen's issues hardly touched the economic grounds of their dilemma." When the Irish-born male operative or laborer, whether Catholic or Protestant, went to the polls, he voted for the Democratic Jacksonian candidate, who appealed to the needs of the newly naturalized citizen, railed against the national bank and paper money, and attacked the basic inequalities of early industrial society.[34]

By 1832 it was clear to the editors of the pro-Jackson *Pennsylvanian* that the Jacksonian platform coincided with the aspirations and ideals of the Irish wage-earners in the city. One week before the national election of 1832, the *Pennsylvanian* addressed the "Working Men of the City and County." It voiced a quite different appeal than the call made by the *Mechanics' Free Press* for the Workingmen's vote:

> To whom would you go for counsel or information in respect to your political and social rights?—to those who . . . live in idle affluence [while] you [live] in unremitting labour . . . ? Will you not depend on those who fought side by side with you in '98?

The *Pennsylvanian* then touched, not on the need for reform in education, the militia law, or the lien law, but on the question of wage reductions:

> Has [your enemy] ever told you that he had reduced your wages, for no other cause than that he would force you and yours to feel want and privation, that he might attribute it to the Tariff, the Veto or something else with which it has nothing to do?[35]

That fall, fearful that the mill town of Manayunk was indeed becoming a "Jacksonville" in their midst, Roxborough's Workingmen's faction aligned with the previously unorganized National Republicans of the township in opposition to the reelection of Jackson. They formed an unprecedented union of farmers, Wissahickon merchant millers, and tradesmen. Nine of the fifty-one individuals who constituted a committee of vigilance to oppose the Jacksonians in September had been active in the committees of the Workingmen's Party. The ranks of the conservative coalition were filled out by men similar in background—property-holding skilled craftsmen. Allying with these tradesmen were the old economic elite of Roxborough: the Rittenhouse millers and the wealthy farmers, including Amos Jones, George Kelly, and George Moyer. Moyer's son John, who was accumulating wealth through land development in Manayunk, joined this political organization that put forward the venerable community patriarchs Nicholas Rittenhouse and Charles Levering for inspector and assessor.[36]

It was not so much the issues of the day—the bank veto, protective tariff, and internal improvements—that brought Roxborough landowners and flour millers together with petty producers and tradesmen in political alliance as it was a mutual fear of the industrial social order. They found clear evidence of the inimical developments of factory society in the influx of laboring-class immigrants, the transparent exploitation of them in the mills, and the extremes of poverty and economic power in the "Jacksonville" mill town. The mills of Manayunk, like those of Manchester, "embodied a double threat to the settled order": on the one hand stood the new economically powerful, politically influential industrial capitalists; and on the other ranged an impoverished, transient, largely immigrant proletariat.[37] Much of the danger lay in the conflict-ridden employer-employee relationship within the mill, from which the traditional modes of social control had been stripped. Writing as "Plain Truth" about the "history of the manufactures of this place," Charles Hagner, in August 1837, brooded over the disturbing nature of the new social relations. Referring to the effects of wage reductions and the subsequent turnouts, he wrote:

> These continual difficulties between the employer and employed, destroy all good feeling and affection between them; the employer generally speaking, cares little for them or their welfare; and they have little regard for him or his interests; he contracts with them to do a certain amount of work; they do it to the letter of the agreement—not one jot more. On pay day he pays them to the last cent, and here ends all communication between them. Instead of the good feeling that ought to exist... the very reverse is too often the case, and I have known many instances where the bitterest animosity existed.[38]

Those in the mill town and up on the ridge surely believed that the conflict between the new industrial employer and employee would not be contained within the factory gates.

Those who formed the hybrid Workingmen's-Whig Party of Roxborough hoped that social and economic reforms such as public education, a mechanic's lien law, and abolishment of imprisonment for debt would bring social and cultural cohesion to early industrial society. The anti-Jackson, pro-reform ticket was victorious at the Roxborough polls as it was throughout the county in October 1832. Yet the fervor of party politics faded with the coming of winter, to be superseded for the moment by what began as an evangelical attack by established tradesmen and shopkeepers against what they perceived as the causes of contension within their society. In the winter of 1832–33, revival meetings swept many laboring-class women and men into the Baptist and Presbyterian churches. These churches were dominated, on the one hand, by petty proprietors and manufacturers and, on the other, by a textile mill-owning family. When placed within the sequence of political and strike activities in Manayunk, the revival seems to have served as a source of political awakening for both the bourgeois and working classes.

The revivals that broke out in the Roxborough Baptist and First Presbyterian churches appear as a moment of phenomenal religious awakening in Roxborough. The growth of church membership rolls during the mass conversions alone testify to this.[39] In the fall of 1832, the newly established First Presbyterian Church of Manayunk was composed of 16 members. At the end of 1833, the church consisted of 84 members, an impressive fivefold increase in one year. The rolls of the Roxborough Baptist Church continued to swell. By the end of 1833, 224 names appeared on the old Roxborough church's membership list, a jump of over 200 percent in two years in the forty-three-year-old congregation.

The letters of correspondence from the Roxborough Baptist patriarchs to the Philadelphia Baptist Association document the evangelical fervor that overtook the mill town in the winter of 1832–33, following the anti-Jacksonian political party victory in the fall. In October, one year after the arrival of Dyer A. Nichols as their pastor, the Roxborough Baptists reported that protracted meetings and frequent assembling for prayer had taken place in the church that stood far up the ridge from the mills. Fifty-six new converts had accepted God's salvation and joined the Baptist fold. Between the following November and February, however, in "the most powerful work of grace known since the first establishment of the church," 109 women and men were baptized by Nichols. One of the conversion meetings of the winter stretched for twelve days. Nichols described the excitement that enveloped the congregation in these months:

> The place of the meeting became the gate of heaven—the number that felt the power of God was large. . . . The means used in this good work were faithful preaching, prayer-meetings, and conversations with anxious persons, calculated to fix their minds on Christ.[40]

Nichols was not, according to one of his successors, a professional evangelist, but a preacher of exceptional talent and energy, assisted in his work by other pastors and licentiates. The pastor of Roxborough Baptist Church in the 1880s recalled that Nichols introduced a "new order of things . . . long protracted stirring and exciting preaching, attractive singing, the anxious seat—in a word the new revival measures of the day."[41]

What was the significance of this evangelical enthusiasm in early industrial Roxborough? How was it a part of the process of the shaping of a new social order? American historians of the past generation have shown that early nineteenth-century evangelicalism struck equally hard in interior manufacturing centers and entrepôts, old port cities, and commercial agricultural regions of the Northeast. Studying communities in different stages of economic development, they have identified various mainsprings of revivalism. All have argued, however, that evangelical Christianity was perceived by members of the emerging bourgeoisie as a means of stabilizing the social order in a volatile period of expanding capitalist relations. Thus in Rochester, in the countryside surrounding Utica, and in Rockdale, Pennsylvania revivalism functioned to promote concord and conformity. The social relations of early industrial capitalism, however, dominated none of these societies as they did Roxborough and Manayunk. And it is here that we find a more complex picture of the significance and meaning of revivalism. "The use and meaning of religious values," Bruce Tucker has noted in a discussion of evangelical Christianity, "is class specific."[42] Indeed, in early industrial Roxborough, the emerging industrial working and capitalist classes fashioned separate and even opposing uses and meanings out of the Baptist and Presbyterian awakenings.

On the eve of the revivals, Roxborough's Baptist church was guided by the propertied tradesmen who had deep roots in the township. Perry Levering and William Rawley, who had joined the church in a small revival in 1821, before the arrival of the textile mills, became deacons in 1830. Both appeared as carpenters on the tax lists of the period and both profited as landowners and developers from Manayunk's industrial development. A fellow convert of 1821, the skilled machine-maker Beriah Willis was also appointed deacon in 1830. Two other deacons, Charles Levering and Benjamin Miles, had joined the church relatively late, transferring in 1828 and 1831 from one of Philadelphia's Baptist congregations. Both, however, had been longtime inhabitants of Manayunk. At the same time that they chose to join the Baptist congregation on the ridge, where they quickly rose to positions of authority, Levering and Miles became active in local anti-Jackson politics and the campaign for democratic reforms. They constituted the Baptist patriarchy that called to the pulpit the dynamic Dyer Nichols who conducted the evangelical revivals of 1832–33.[43]

The course of revivals in the Roxborough Baptist Church resembled the two-stage process that occurred in Rochester, New York, from 1832 to 1836. There, Paul Johnson has argued, tradesmen-employers, converted in an earlier revival, led a second awakening in order to bring members of the wage-laboring population under the moral precepts of evangelical Christianity. Religious enthusiasm struck first among Rochester's employers and petty manufacturers in the 1820s and then spread, through the efforts of the early converts, to the ranks of the laboring classes. This sequence of conversion also occurred in the Roxborough Baptist Church, even though a decade of stagnation separated the small but significant employers' revival of 1821 and Nichols's working-class revival of 1832–33. Along with carpenters, masons, blacksmiths, and machinemakers, laborers and mill workers came into the church during Nichols's revivals. The old, propertied families such as the Leverings, like their counterparts in Rochester, must have perceived evangelicalism as a force of cohesion, for all who joined the Baptist fold were exposed to the moral injunctions that hard work, honesty, and submissiveness were continuing signs of grace. Such evangelical teachings dovetailed neatly with the ideology of social reform. Men like Levering and Miles became involved in reform politics and evangelical religion because they shared a concern about the problems, as Johnson has put it, "of class, legitimacy, and order generated in the early stages of manufacturing."[44]

The revival that took place in James Darrach's Presbyterian congregation in the heart of Manayunk paralleled the Baptist revival. Before the fall of 1832, the First Presbyterian Church of Manayunk (formerly the Dutch Reformed Church of Manayunk) consisted of sixteen members. Headed by Cordelia and James Darrach, the textile mill owner, the congregation accepted a young minister from the New School Fifth Presbyterian Church of Philadelphia, where the evangelical minister Thomas Skinner presided. In October, he began conducting prayer meetings in the stone building connected to Darrach's mill. No descriptions exist of this revival as they do for the Baptist church. The success of the revival in the small church, however, is borne out by reports that seventy-nine students attended a new Sabbath school that met in the mill building. Membership lists show that a year after the arrival of the New School minister, the congregation had jumped to eighty-four members.[45]

James Darrach certainly must have perceived a positive relationship between evangelical conversion and an industrious and submissive work force. Implicit within the doctrine of individual salvation was the consensual notion that "economic conditions were a matter of fate, rather than for struggle."[46] Darrach experienced more immediately than the Baptist patriarchs the problems of legitimacy and control in a period of tension between factory workers and employers in Manayunk. He was no doubt attracted to the thrust of New School Presbyterianism, in part because of its potential for dissipating an oppositional stance among the mill workers.

Who among the wage-laboring class participated in the Baptist and Presbyterian revivals, and what meaning and message did they derive from evangelical religion? Twice as many females as males, the same proportion that existed in the mill labor force, were baptized in both the Baptist and Presbyterian churches. The majority of these women who flocked to Roxborough's Baptist revival, as well as to the Presbyterian revival, probably were not wives and mothers from middle-class families but employees of the textile mills. No mill records exist to link with certainty female operatives to the church membership records. However, the records identify three male mill workers and their families who joined the Presbyterian church in the spring of 1833. And we can glean from the information in the Baptist church lists evidence to suggest that geographically mobile female mill laborers were an important part of the mass conversions. In November 1831, Elizabeth Patterson was baptized into the church on the ridge in one of Nichols's earliest revivals. No male bearing her surname appeared in the records. However, on January 6, 1833—at the height of the evangelical excitement—Mary Patterson was baptized along with sixteen other converts. She was probably the daughter of Elizabeth, and both, it would seem, held jobs in one of the textile mills on the canal. On March 15, 1834, a few days after a handful of Manayunk mill owners announced wage reductions and some mills closed, Elizabeth and Mary Patterson asked for and received a letter of dismissal to the Baptist church in Norristown, the large mill town ten miles up the Schuylkill from Manayunk. In August 1840, the mother and daughter were received back into the Roxborough congregation, no doubt returning to find employment again in the textile town they had left six years before.[47]

A significant number of the Baptist revival participants migrated to Norristown within months and sometimes weeks of being baptized. On two successive Sundays in late November 1832, ten females and two males were baptized along with dozens of others. On December 9, within three weeks of participating in the mass baptisms, these twelve obtained letters of dismissal to Norristown. Like the Pattersons, these women and men had probably worked in the mills below the church on the ridge. Five of the females shared the same surname. It is likely that one was the widowed mother of the others, and they were on the move from one textile town to another.[48]

Manayunk's mill workers undoubtedly sought a specific meaning in the evangelical meetings. We have seen that women and men labored difficult hours in the new mills, earned a bare sufficiency, and seemingly pulled up roots with the frequency of the seasons and the industry's recessions. It makes sense that they would be attracted to the vital evangelical message of spiritual peace and equality and to the fervent experience of mass conversion. As Eric Hobsbawm has argued, the evangelical promise of hope and salvation was pertinent to the new industrial workers for whom life was "miserable, poor, nasty, brutish, short and above all, insecure." Hobsbawm has suggested how it may also have defused class conflict. The mill worker's vision of judgment and hellfire for the employer who exploited his workers in this life may have helped her to bear the burden of her temporal struggle. It is impossible to tell to what extent this form of "psychic satisfaction," as E. P. Thompson has called it, inhibited class consciousness among the mill workers.[49] The act of participating in the evangelical revivals, however, may have been one of political, as well as religious awakening, particularly for female operatives, for it gave those who were generally mute the courage to speak out and those who were typically uprooted a sense of community.[50]

The course of events that followed the period of revivals in Manayunk suggest that the working-class participants did not accept the ideology of passivity and of the sanctification of hard work as their employers hoped, but drew their own uses and meaning from the revival experience. No sooner had the revivals faded in the early spring of 1833, when mill operatives and laboring people of all denominations began a series of strikes that would last for two years. For those employers who had hoped otherwise, the onset of the strikes quickly demonstrated that neither the evangelical doctrine of resignation, patience, and submission nor the anti-Jacksonian agenda of democratic reforms could resolve the conflicts and tensions within early industrial society.

VII

THE MILL STRIKES AND THE NEW IDEOLOGY OF CLASS

During his visit to the eastern seaboard in 1831 and 1832, Alexis de Tocqueville was struck by the coercive power of the new industrial capitalists. His depiction of the danger they posed for democratic society captures the particular tension that permeated Roxborough society on the eve of the Panic of 1837 and the six years of depression that followed it. De Tocqueville was of the opinion "that the manufacturing aristocracy which is growing up under our eyes is one of the harshest which ever existed in the world." While the French observer acknowledged that it was "one of the most confined and least dangerous," he warned the keepers of the Republic to watch the manufacturing class closely:

> The friends of democracy should keep their eyes anxiously in this direction; for if ever a permanent inequality of conditions and aristocracy again penetrate into the world, it may be predicted that this is the gate by which they will enter.[1]

Indeed, in the years between 1833 and 1837, de Tocqueville's predictions gradually materialized as a "manufacturing aristocracy" and an industrial working class began to take form around the mill gates of Manayunk. Class conflict and consciousness sharpened through a series of events that unfolded between 1833 and the Panic: first, the strikes and union movement of mill workers in 1833 and 1834, which duplicated events unfolding in Manchester; second, the expansion of the working-class base of the Jacksonian Party and birth of the Whig Party in 1834; third, the rise of the temperance movement and the outbreak of a general strike in 1835; and finally, the division of both the Jacksonian and Whig parties into warring factions in 1835 and 1836. These events came to a head on the eve of the Panic of 1837, when all social groups in Roxborough came to recognize the industrial mill owners as a capitalist class whose attitudes and interests were in opposition to their own.[2] It was the mill strikes of 1833 and

1834, however, that proved to be the decisive juncture in the emergence of an early industrial working-class consciousness.

A central factor in these events was the immigrant textile workers' critique of early industrial production and their role in the broad-based labor movement of the early 1830s. Unlike their contemporary counterparts in Lowell, Lynn, or New York, Manayunk's labor militants and trade unionists brought an imported body of ideas to bear on their American industrial experience. The strike leaders and spokesmen drew on the oppositional tradition of England's anti-Ricardian economic theorists and radical trade unionists who cogently set out an attack on the developing capitalist system in the years following the appearance of David Ricardo's *Principles of Political Economy* (1817), a classical economist's interpretation of capitalist relations. The single most important influence on the dynamics of class conflict in early industrial Manayunk was the transmission of trade union ideology from the northern textile districts of England by Manayunk's immigrant mule spinners and powerloom weavers.[3] It intertwined with America's indigenous "artisan republicanism"—the belief in the preeminent status of the producer vis-à-vis the nonproducer and the idealization of the community of independent and virtuous craftsmen—to create a transatlantic laboring-class ideology.

In 1832 American textile manufacturers were facing rising prices for raw cotton owing to an increased demand in England for the Southern commodity. By the beginning of the following year, those Philadelphia mill owners who were unable to maintain production under conditions of increased costs and a competitive and recession-prone market were discharging employees. In the early days of August 1833, Joseph Ripka and J. J. Borie, Manayunk's two largest employers, utilized another tactic. They announced a 20 percent reduction in the wage rates of all their mill operatives. It was the first major wage cut since late 1828, the cause of the last industrial strike in the mill town.[4]

On Wednesday, August 7, the *Germantown Telegraph* revealed that "some little excitement" existed between the workmen employed in several Manayunk factories and their employers over the proposed wage reductions. On the following Monday, operatives from Ripka's factories met outside one of Manayunk's inns and formed themselves into an organization they dubbed "the Working People of Manayunk." Their stated purpose was "to propose and adopt measures for a steady stand against the attempt of Mr. Ripka [to cut wages]." In two days an acting committee of seventeen cotton spinners, machinists, shopkeepers, and tradesmen had drafted resolutions expressing the sense of the meeting and read their report before the second gathering of the mill operatives.

The resolutions were published in the *Germantown Telegraph* on August 14, 1833. In the preamble, the workers declared that a reduction would "bring many of us to poverty and distress." Since they had no other means of "gaining a livelihood except by working in the factories," the operatives had no choice but to resist wage cuts that would reduce "many families to a low condition." In this first public pronouncement by the strike committee, the workers stressed repeatedly that if it were not for the fact that their wages were already so low and that their long hours precluded supplementing their income with other jobs, they might be able to accept the present reduction. Needing the support of the "generous public . . . to bear us successfully through the contest," the operatives portrayed themselves as reasonable men and women, wanting only what justice demanded—that their earnings not drop below "a bare sufficiency." They called on the producers of the community, "the working people of all Trades and callings," to give them assistance.

If the editorial remarks of the *Germantown Telegraph* and *Pennsylvanian* can be taken as indicative of public sentiment, the "generous public" initially sympathized with the workers' cause. On August 7, in cautious support of the Working People of Manayunk, the *Germantown Telegraph* wrote, "We are sure that public opinion will sustain the honest industrious manufacturers, in persisting in a rate of wages, which as far as our knowledge extends, is barely sufficient to the economical and decent support of themselves and families." Ten days later, the pro-Jackson *Pennsylvanian* took the opportunity to attack the protective tariff of 1832. That tariff, the editors argued, had allowed capitalists, who were now reducing wages, to "get fat on protection." Who can blame these working people, the editors asked, if they "rebel under such a system?"[5]

The mixed composition of the committee that drafted the Working People's resolution of August 8 indicates that prior to the turnout, broad-based support for the mill workers' cause existed among Roxborough's laboring men and proprietors, many of whose wives and daughters probably worked in the mills. The fact that future labor leaders from Philadelphia were on the committee further suggests the importance of the factory operatives' strike to the broader cause of a burgeoning trade union movement in that city. It did not take long for the conflict between mill worker and owner, as articulated by the standing committee, to move quickly beyond the issue of the wage reduction and the worker's family economy.

Seventeen men served on the acting committee that composed the strike resolutions. Three can be identified as mule spinners, members of a trade that spearheaded the industrial labor movement in Britain.

The Mill Strikes and the New Ideology of Class

William Crook knew firsthand the determination of Manayunk's mill owners, for he had been involved in the strike against Borie in 1828 and had testified on behalf of those mules spinners who had been brought to trial for conspiracy.[6] James France, also a mule spinner, had been born in Manchester in 1789. At the age of forty, he and his family immigrated to Manayunk. As a Manchester mule spinner, France had been at the center of industrial labor's organized strike and union movement. By 1810 spinners throughout Lancashire were organized in trade clubs. And in that year, they struck to raise wages in the smaller centers surrounding Manchester. France, a young man of twenty-one, likely participated in what was called as late as 1834, "the most extensive and persevering strike that has ever taken place."[7] Spinners and powerloom weavers struck again in 1818 to recover some part of the wage reductions they had suffered since 1810. And in the beginning of 1829, around the time that France and his family left Manchester, the spinners embarked upon a year of struggles against a series of wage reductions. France no doubt brought to the textile workers' committee in Manayunk valuable knowledge and experience of the opposition of Lancashire mule spinners against wage reductions.

The composition of the committee cut across cultural and occupational groups. Jacob Kuhn, a machinemaker who worked in the mills, had been converted in the revivals of the previous winter and baptized into the Roxborough Baptist Church. His participation raises two important historical points about the relationship of evangelical religion and class identity. First, as a laboring man his own experience of conversion and salvation did not translate into quiet acceptance of the new industrial order and the unchallenged authority of the capitalist employer. The culture of "the revivalist" of the working class seems more variegated than Bruce Laurie has suggested. Second, the political demands of evangelical Protestantism did not yet divide Protestant workers from Catholic as was true in Philadelphia a decade later. Irish-born Catholics Thomas Mullin and Patrick McGlinchy were appointed to help draft the resolutions at the meeting of August 8. They were both among the twenty founding members of Manayunk's Catholic congregation organized and overseen by Jerome Keating. Their role in organizing the strike indicates that Keating's authority as Catholic patriarch was limited. Moreover, neither of them labored in the mills; McGlinchy was a storekeeper and Mullin a gatekeeper on the canal. Their representation on the committee suggests that producers and proprietors throughout the wider community aligned with the mill workers' cause.[8]

Experienced Philadelphia labor activists involved in the political and trade union movement among the city's skilled trades immediately joined the committee and lent the strike a wider political content. William Gilmore, a shoemaker from the Southwark district of Philadelphia, attended the first organizing meeting. His role in the Manayunk strike would propel him into the leadership of Philadelphia's trade union movement in the mid-1830s. Gilmore probably contacted his friend and fellow labor activist John Ferral, a handloom weaver at the Falls of Schuylkill, for he was added to the acting committee. They would soon rise together as luminaries in Philadelphia's and the nation's insurgent labor movement. Undoubtedly, both men judged the factory workers' cause to be a unifying force for the skilled trades of the city. One other Philadelphian joined the original acting committee, and he was the only member that can be identified outside of the laboring classes. William F. Small, a lawyer and radical, was perhaps enlisted to lend a practiced hand in composing the public appeals of the striking mill workers.[9]

The resolutions of Ripka's striking employees, penned by this coalition of laborers, proprietors, and radicals, galvanized operatives of other mills where wage reductions had been imposed. Within a week of their public reading, two to three hundred hands from Borie and Keating's Schuylkill Factory (popularly known as "Mr. Keating's Mill"), along with the operatives from two other small canal factories, turned out to resist the wage reductions. Inspired by the widespread protest among Manayunk's mill operatives and by the general air of industrial crisis, the Manayunk Working People's committee met on August 23 to read a lengthy public proclamation urging the textile laborers to resist the wage reductions. Gilmore and Small presided over the committee and used the proclamation to call for a union of all trades to oppose "tyranny" and "an overbearing aristocracy."[10]

The critique in the "grievances and petition" of the Working People of Manayunk combined a moralistic natural rights concept of exploitation with an economic analysis of the injustices of competitive capitalism. The petition began by exposing the conditions of work: "Thirteen hours of hard labor ... in an atmosphere thick with the dust and small particles of cotton." The entire family, the workers noted, had to labor in this debilitating atmosphere because wages were "barely sufficient to supply us with the necessaries of life." The authors called attention to factory reform in England, arguing that the abuses of child labor had been corrected in English mills while Manayunk children were sent to the mills at an early age, there to be "reared in total ignorance."[11]

Turning to the issue at hand, the petitioners criticized the employers who reduced their meager wages by 20 percent because the price of cotton had risen. Here the logic of a moral economy was invoked against the workings of a competitive system. "What matters to us what the price of cotton is[;] our wants are as great when cotton is dear as they are when it is cheap." In the same breath, they condemned the appropriation of their unpaid labor, noting that they did not receive better wages when their employers' profits went up. The powerful specter of slavery was ever-present in this nascent working-class consciousness. The committee left no doubt that they were waging a struggle of an oppressed class. To submit to the wage reduction, they stated, would be to "rivet our chains still closer."

> We have long suffered the evils of being divided in our sentiments, but the universal oppression that we now all feel, have roused us to a sense of our oppressed conditions and we are now determined to be oppressed no longer![12]

For any who might question the centrality of the issue of wage reductions in the conflict between labor and capital, the Working People's petition reminded them that it was but the "forerunner of greater evils, and greater oppressions, which would terminate, if not resisted, in slavery."[13]

The public proclamation of the Working People of Manayunk revealed a sharpening sense of working-class identity and unity. Imbedded in their rhetoric were republican sentiments that inspired their condemnation of the system of competitive capitalism. The capitalist mill owner was repudiated for belonging to the class of "overbearing aristocracy," while the industrious worker and the poor were becoming one and the same. But the argument did not culminate in a call for the restoration of a harmonious past where equality of status prevailed among producers. While "divided in their sentiment" before, the workers acknowledged that they were now aware of their oppression as a class. Such ideas, which juxtaposed capitalist employer and laborer as exploiter and exploited, filled the trade union and cooperative literature of Lancashire in this period. The particular nature of the system of production in Manayunk made them pertinent.[14]

While the language of class conflict may have become commonplace in Manchester, it jolted the community surrounding Manayunk's mills. The militant proclamations that publicized the workers' growing frustration with conditions in the mills made the public increasingly aware of the potential for violent conflict. On August 28 the *Germantown Telegraph* reiterated its opposition to the wage reductions but urged the workers to act sensibly. "There is not perhaps an individual among us, but who is fully aware of the advantages to the community, from the allowance of a reasonable rate of wages."

The strikers, however, should pursue "a prudent and praiseworthy course," counseled the editor, or risk losing public support. Their proclamation had conjured up a frightening specter for some in the community. On August 31, two days after the radical proclamation appeared on its pages, the *Pennsylvanian* reported that "many talk about the atrocity of combinations among the operatives."

While the committee of radical operatives and tradesmen appealed for support of the standout, the striking female powerloom weavers, male mule spinners, and child piecers carried out the resistance in street rallies and picket lines. Their action, in conjunction with the radical manifesto, left little doubt of the antagonistic sentiments of the industrial laborers. Charles Hagner characterized the organized turnouts of the early thirties in Manayunk as "bad schools" for the young workers. He seemed uncomfortable, as well as unfamiliar, with what he observed. "They early acquire the views and feelings of some of the older hands; and seem to me, in some instances, to inherit an antipathy to their employers, which grows with their growth and strengthens with their strength." Hagner suggested that hostility between employers and employees did not evaporate with the end of the strikes of the early 1830s. Antagonisms were transferred from older to younger workers as time passed. Women, he went on to point out, often led young girls in the demonstrations.

> On the occasions of these "strikes" it is not an uncommon thing for the women and girls to assemble on the commons in public meeting, pass resolutions, and listen to the harangue of someone, the burthen of whose song is abuse of their unfortunate employer.[15]

Such independent organized action by women operatives was common within the British textile trades. Female spinners and weavers contributed to the great surge or working-class protest and opposition in the late 1820s and early 1830s. At Wakefield in 1833, magistrates charged female spinners with harrassing and assaulting women who were "blackies," that is, black sheep who did not join the union. Indeed, in the same year that Manayunk female powerloom weavers joined the district's first strike movement, female weavers and spinners in Glasgow combined in an effort to raise wages to the same rate as that paid to men.[16]

Evidence that could shed additional light on the thoughts and actions of women in the Manayunk strike movement and on the dynamics between male and female participants is lacking. However, Hagner's account of the strikes indicates that women and children of the factories—the majority of the operatives—participated actively in promoting their interest in the one sphere that was not closed to them: the public demonstration.[17] Male mule spinners also took to the streets. On one occasion, they threatened a traditional laboring-class justice. When their boss spinner Tony Kerns sided with mill-owner Ripka, the striking operatives paraded down Main Street singing:[18]

> Old Tony, he owns a house
> Old Tony, he owns land,
> As far as we can understand
> He's at Joe Ripka's command,
> Oh rise up ye spinners!
> Don't let your courage fail,
> If Tony Kerns goes into work
> We'll ride him on a rail!

While mule spinners, their child assistants, and female powerloom weavers and throstle tenders kept alive the strike movement through public meetings, discussions, and demonstrations, a battle raged in the pages of the *Germantown Telegraph* between spokesmen for the mill owners and for the mill workers. For three months, from the end of August to the end of November (when the editor stopped printing the letters), "A Jeffersonian Workingman," "Perceiver," and "J. F." (John Ferral) presented the grievances of the industrial working class. Their views were challenged by the unwavering defense of "An Observer," who upheld the rights of the capitalist mill owners.

The debate strengthened support for the strike. Observer revealed this when he blamed Perceiver for "exciting [the mill workers'] passions and prejudices against their employers, and preventing an arrangement between them." In a letter of November 6, Observer held Perceiver partly accountable for three of the six cotton mills in Manayunk being entirely stopped by the beginning of November. Like the strikers' proclamation to the public on August 23, penned by trade unionists Gilmore and Ferral, the anticapitalist arguments posed by the correspondents broadened the political content of the "defensive" strike. The letters in the *Germantown Telegraph* linked wage reductions to the contradictions of early industrial capitalism, making the community aware that the workers' grievances about wages could not be divorced from the broader conflict between owners and workers. Throughout the debate, the hostile and provocative language slackened only when the opposing parties marshalled evidence and statistics on the price of cotton, work rules, wage rates, and the economic impact of the tariff, banks, and militia law upon mill workers.[19] In the argument posed by the workers' advocates, we can observe the important influence of Lancashire labor radicalism on the oppositional thought of Manayunk's factory workers.

The debate began on August 21 when "A Workingman" penned a letter to the *Germantown Telegraph* detailing many of the grievances that would appear in the proclamation read before the meeting of the Working People of Manayunk two days later. Perhaps a member of the strike committee, the author displayed an insider's knowledge about the conditions and wage rates in the canal factories. He pointed out that in Ripka's factory the best mule spinner, who worked thirteen to fifteen hours per day, could earn at most $4.33 per week. Using language that the editor characterized as "some rather harsh expression," Workingman attacked the cutting of such wage rates as a mechanism that perpetuated a system of economic inequality. The notion that "the rich and the capitalist were one and the same" was imbedded in the Workingman's critique: "It is plain as the sun at noon-day, that as the poor are sinking, the rich are rising." The proposed wage cuts, he asserted, would "increase the poverty of the poor, and add to the riches of the rich."[20] Workingman's criticism of the dynamics of the early industrial economy, like that of the Working People of Manayunk, was grounded partly in a philosophy of natural rights. Invoking the revered sentiments of the revolutionary generation, he wrote, "We are all born free and equal; it was never intended that the employer should take the place of the master, and the employed that of slave." Ancient liberties were at risk. "Are we so far debased," he appealed on September 8, "as to be afraid to assert our rights, the rights of freemen, to break the shackles of oppression?" As Workingman had warned, the factory spinner now stood to lose his economic independence, for he "could scarcely hope to become his own master." Nevertheless, his political rights must not be impaired.[21]

Two weeks following this assertion, "A Workingman" had tactfully become "A Jeffersonian Workingman" and more explicitly linked the current cause of the Manayunk mill workers to that of the American Revolution. If workingmen did not overturn the class-biased laws that created banks, tariffs, and the militia system and allowed capitalists to accumulate labor-saving machinery and mills, he argued, then "we will have to enact the scenes of '76 over again, to get rid of our internal tyrants."[22]

Workingman's view of the capitalist system was also informed by the labor theory of value and the correlative idea that society had become divided between a producer and a parasitic nonproducer class—a line of laboring-class thought nurtured in England for more than a generation and central to the mechanic's ideology of the Jacksonian era. The textile operatives were robbed of the value of their labor,

Workingman argued, by those who "produce not, and yet engross the produce of others."[23]

A nostalgic view of the wrongs suffered by the independent producer expressed by Workingman soon gave way to a more radical critique drawn from England's anticapitalist theorists and labor leaders. In the 1820s, critics of political economist David Ricardo laid out the theoretical basis upon which British labor leaders would build their attack of the early industrial capitalist system. The radical economists used a proto-Marxian analysis of surplus value to dispute conservative political economists who asserted that labor was a commodity and the product of that labor belonged to the owner of the means of production. A product derived all its value from labor, the radicals argued, and therefore the worker was entitled to the whole of what he produced. The basis of injustice, then, lay in the appropriation by the capitalist employer of the surplus value of labor in the form of profits. The cotton spinners' leader John Doherty, who was familiar with the writings of anti-Ricardians William Thompson and Thomas Hodgkins, was expressing an explicitly anticapitalist view when he championed the trade union to stop the nonproducers from "robbing the working people of their right to the whole of what they produced."[24]

In the transmutation of the pro-labor newspaper correspondent from "A Workingman" to "A Jeffersonian Workingman" and finally to "A Jeffersonian American Workingman," the influence of the English radical analysis became evident. Indeed, the importation of the anticapitalist ideas of northern England's working-class movement became a central issue of the three-month debate. After Workingman's first letter appeared, the editor of the *Germantown Telegraph* himself politely advised him to use "a little more moderation" in his communications if he wanted to influence favorably the "very respectable class of persons" in the community. The editor became silent, however, as "An Observer," solicited by the mill owners, took Workingman to task.[25] On September 4, Observer marked Workingman as a product of foreign radicalism, a "conceited visionary theorist" under the influence of "rooted prejudices acquired in *other countries*."

Attempting to undermine the broad-based support enjoyed by the mill hands, the mill owners' spokesman appealed to the common sense and nativist sentiments of Roxborough's tradesmen. He assured his audience that no matter how those "petty reformers" with "wild and extravagant theories" might have "figured in the vicinity of Manchester," they were "Fish out of Water" in Manayunk. The wage reduction imposed by the mill owners was justified by the "plain simple facts of the case," Observer argued, most particularly by a 50 percent increase in the price of cotton in one year. Furthermore, he argued, the mill owner, as part of "a business agreement," had the right of contract to lay off or reduce the wages of his employee. The employee was "at perfect liberty to accept or reject" the reduction. If it did not suit her, the operative could seek a better offer elsewhere. Workingman had argued with "more zeal than knowledge," Observer ridiculed in the letter of September 4, and knew as much about the matters of wages and mill contracts "as the man in the moon."

Two weeks after Observer's letter was printed, the "Jeffersonian American Workingman" appeared to defend the views of English labor leaders on political economy. He insisted that such ideas deserved serious attention and articulated the "plain simple facts" of the relationship between Manayunk and Manchester. Manayunk did indeed resemble Manchester in the eyes of working people who had experienced the "abject misery" of industrial England. It should surprise no one, he exclaimed, that upon "seeing the same system in active operation here," the immigrant workers expressed "their disappointment in language grating to our ears."

Observer's ridicule of Workingman for bearing prejudices from other countries prompted "Perceiver" to join the debate on October 9. He attacked Observer for his insolent nativism and revealed the burgeoning ideology of a transatlantic industrial working class:

> And do you, if a born native of these states, who never resided in any other country, really think yourself capable to give advice to foreigners, who have worked in factories there and here, about their rights as working people?

Observer continued to raise the specter of foreign radicalism in further correspondence. In a letter that appeared on October 30, he appealed to "the honest working-people of Manayunk," proclaiming that Workingman was not a "True-born American" and ought to cut his connections with the early socialist "Manchester New Lights." When Perceiver responded on November 6 that it was the fact of the matter that families could not subsist upon their wages in Manayunk, and not what Observer referred to as "the *wild theories of Manchester reformers*" that caused the strike, the mill owners' spokesman shot back with a detailed exposé of the reformers and their ideas. First, the typical Manchester subversive was a recent immigrant to America, having been in the country no more than eighteen months. He was "a rank *Atheist*" whose reading was limited to Fanny Wright, Robert Carlyle ("that great nuisance to society"), Thomas Paine, and Robert Owen. Having been raised "from infancy in a factory," this foreign imposter in the mills, Observer warned, would "talk to you for hours, about liberty, equality, the rights of man, Declaration of Independence, &c." As if offended by the pretensions of a mere immigrant worker to know of such things, Observer lashed out: "Attempt to reason with him—bah! It is literally throwing pearls to swine! Miserable, ignorant, malicious, infatuated wretches, reform *us*! our country and its institutions."[26]

John Ferral, the Philadelphia handloom weaver and a member of the recently formed strike committee, joined the verbal fracas at the end of October. "J. F." expressed as much contempt for Manayunk's mill owners as did Observer for the immigrant mill workers, and he matched his most clever metaphors. The mill owners, he wrote, in a letter of November 13, were "aristo-*rats*." They were "the thieves of the working-men's grainery.... They leave the real owners the chaff; they are the vermin of society... who by means of banks, tariff laws, &c. &c., monopolize *the proceeds* of the industry of the people." J. F.'s message was unambiguous: capitalist institutions deprived the working people of their just claim to their own property—their labor.

Taking Observer to task, Ferral credited the Manchester reformers within English trade societies with having forced into operation the British factory laws to protect workers from the cotton manufacturers—"the speculators in human misery." Following the appearance of J. F.'s correspondence, the nervous editor of the *Telegraph*, announcing that the "political cast" of the letters had offended some of his readers, dropped the debate from the pages of his paper, but not before the community had learned of the deep antagonisms that divided Manayunk's industrial workers and employers, and not before those involved in the controversy had documented the presence of an imported Anglo-American laboring-class radicalism.[27]

In the bold assertions of Workingman (in his various guises), Perceiver, J. F., and their foe Observer is evidence that immigrants did make up the labor force in Manayunk and judged the American industrial world by their experience in English factories. They measured the workplace conditions and system of exploitation, which were as harsh and transparent as any in industrializing America, as mobile and permanent wage-laborers, not as expectant mechanics. While strands of the indigenous artisan ideology appeared in the critique of the mill workers' spokesmen, they fused with a more radical working-class consciousness. Workingman and his allies only momentarily invoked a more harmonious and perfect past and the sacred tradition of revolutionary republicanism.[28] They were quick to proclaim the rights of Anglo-American factory workers in a maturing capitalist system and to drop the "reactionary nostalgia" of the Revolution when Observer dismissed them and their ideas as foreign. The main battle in society was between labor and capital, with control over wages as the central issue.

It was at this time that working people in other industrializing settings in the Northeast were wedding natural rights theory to the labor theory of value to produce a notion of property rights in labor power and wages. The impetus for this emerging working-class ideology has not been located by historians in industrial England, except to the extent that labor leaders were aware of such thought through the reading of the works of England's radical economists and through news of the English working-class movement. Manayunk's strike committee and prolabor newspaper correspondents likely read the published essays of radical economists such as John Gray. His socialist tract *A Lecture on Human Happiness* (1825) was reprinted three times in Philadelphia and appeared serially in the *Mechanics' Free Press* in the late 1820s.[29] Central to labor's awakening in Philadelphia, however, was the transmission of oppositional ideas by Manayunk's immigrant textile laborers themselves, who derived them from long involvement in factory work and protest on both sides of the Atlantic.

The influence of English oppositional thought was given concrete and coherent expression in the trade union movement that was precipitated by the Manayunk mill strikes. Of more immediate concern to the Manayunk mill owners than the "foreigners'" unreserved advocacy of the rights of labor within a competitive capitalist economy or their critique of the system of industrial production in Manayunk was the surge of the trade union movement and, in particular, the establishment of the Trades Union of Pennsylvania. Within days of the formation of the Working People of Manayunk to organize against Ripka, the strike committee considered establishing a permanent and broader organization of factory workers. On 14 August, the neighboring Blockley textile operatives convened in response to the Manayunk strike and made an unprecedented proposal. In the *Pennsylvanian* they thanked "Messrs. Ripka and Co. for awakening the workmen of Manayunk and its vicinity to a sense of their danger" and resolved that a "permanent union be established among [the working classes]."[30] By August 23, the Manayunk Working People's committee was calling for information from trade unions in other cities "concerning their regulations, &c." On the same day that the committee's plea for information appeared in the *Pennsylvanian*, Jeffersonian Workingman exhorted the mill workers to set the example for all trades: "Join yourselves together in one firm body, never to be separated, and hand in hand, go forward till you break those shackles of oppresssion and bring your fellow workingmen on that sure foundation which none dare contend with."[31]

On September 9, in the midst of the strike at the Manayunk canal mills, factory delegates from the textile towns of Blockley, Gulf Mills, Brandywine, Pike Creek, Roseville, Haddington, Haverford, and Norristown converged on Manayunk to establish the Trades Union of Pennsylvania (TUP). The Manayunk working people knew that a regionwide union could strengthen the strikers' position. Once formed, it helped sustain the mill strike through October. The Manayunk strike committee itself was apparently subsumed under the TUP. William Gilmore and William F. Small, the president and secretary of the meetings of the Manayunk Working People, held these same positions of leadership in the new regional organization.[32]

A number of factors suggest that the intellectual roots of the factory-based trade union movement can be traced to industrial England. First, the factory cotton spinners of Lancashire, from where Manayunk's spinners emigrated, were in the vanguard of the English trade union movement. Second, the protection of wage rates, which was the heart of the English unionists' cause, was the organizing element of the Trades Union of Pennsylvania. Other trade union movements in America focused on more traditional concerns until 1835. The ideology espoused in Manayunk, as in the northern industrial towns of England, represented the most class-conscious strand of labor radicalism.[33]

English trade unionism, championing the property rights of working people over their labor power and the fruits of their labor, had fermented in the factories and shops of Manchester and London in the depression years following the Napoleonic Wars. In the late 1820s, labor's union movement expanded as local groups tried to form wider combinations to resist a general tide of wage reductions. Again, Lancashire textile workers were pioneers in the national organization of particular trades. The failure of a six-month strike by Lancashire male spinners in 1829 convinced leaders John Doherty and others that only a combination of local unions could succeed against the concerted efforts of employers to reduce wage rates. That summer the Grand General Union of Cotton Spinners was established. Within a year, John Doherty used this organization as a springboard to establish the first general union of all trades in Great Britain. The National Association for the Protection of Labor drew its vast support from various textile trades, and its express object was to resist wage reductions.[34]

The trade union movement of the northern English textile workers derived from the most fully developed working-class critique found in anti-Ricardian thought. In the 1820s, the radical economist Thomas Hodgkins articulated a doctrine of class struggle and advocated that the trade union movement was the best means of organizing the working classes to resist capitalist exploitation. John Doherty constantly expressed anticapitalist views in his journals and pamphlets. He and other trade unionists rejected Owenite cooperativism in this period. There is no evidence that the Manchester spinners' society or the spinners' general union became involved in cooperative production or selling. Nor is there evidence that Manayunk workers did. Manayunk operatives, like Manchester factory workers, put their faith in the trade union to resist the wage reductions of hostile employers and to oppose the political economy of the early industrial capitalists.[35]

It was the class-conscious Lancashire-based movement among England's industrial workers that Joseph Ripka explicitly blamed for the mill strikes in Manayunk. When Ripka testified before the Pennsylvania factory investigation committee in 1837, he pointed to the transmission of the Anglo-Irish "principle of the Trades' union, which has been introduced amongst the laboring classes in general," as the "greatest evil" of the factory system. From the capitalist's point of view, it had no basis in American condition or custom. Trade unionism, Ripka asserted "has been imported to this country by English and Irish men within a few years." It was these foreign ideas, Ripka believed, that lay at the base of the strikes, for they destroyed "the good feeling which has, heretofore, existed in this country."[36]

Following in the footsteps of Manchester's Doherty, William Gilmore moved to broaden the working-class base of the industrial union of textile operatives. In early December 1833, he called delegates of "Farmers, mechanics and Working-men generally" from throughout the state to a convention in Philadelphia. However, the simultaneous rise of the more moderate Trades Union of the City and County of Philadelphia (TUCCP) momentarily divided the trade union movement. The division points up the radical sensibilities of the industrial workers concerning the divisions and alliances that permeated their society. Artisan tradesmen, not factory operatives, gave birth to the TUCCP. Inspired by the meeting that fall of the craft-oriented "Mechanics, Farmers, and the Working People of the New England States," remnants from Philadelphia's defunct Mechanics' Union of Trade Associations founded the countywide TUCCP. In Manayunk on November 11, 1833, James Renshaw, the anti-Jacksonian innkeeper and former member of Manayunk's extinct Workingmen's Party, hosted a general meeting of the "Workingmen of Manayunk and Vicinity." The meeting of Manayunk's established tradesmen heard a report from their delegate who had attended the prototype convention of the New England mechanics.[37]

Probably few, if any, of Manayunk's factory operatives attended the meeting at Renshaw's Inn, just as an insignificant number of industrial workers had attended the parent conventions in either Boston or Philadelphia. Textile laborers held relatively little interest in the primary goal of the journeymen movements that were arising throughout the eastern cities in 1833: a general system of education that included both public and manual labor schools. The Trades Union of Pennsylvania, in fact, made the decision to exclude education from their official platform. The committee that drew up the convention's platform in December, chaired by John Ferral, recommended only two objects to be pursued by the mill workers' union: the ten-hour system of labor and adequate wages. Like the Lancashire trade unionists, the TUP kept their goals riveted to the class-specific interests of the wage earner.[38]

As factory workers set about creating an effective union of wage earners, employers throughout the county organized themselves in the face of a deepening recession and rising working-class opposition. In December 1833, employers in the city and county united under the designation "Traders, Manufacturers, Mechanics, Merchants, and other Citizens" to consider "the existing pecuniary pressure pervading all classes of the community." They, no doubt, pondered the concerted effort by the laboring classes in the port city to resist the wage reductions that swept through the recession-ridden trades. Two of Manayunk's textile capitalists, Mark Richards and T. B. Darrach, were in attendance.[39]

The exact fate of the Trades Union of Pennsylvania and of the operative's strike in Manayunk that begat the industrial union is regrettably unclear. The TUP met for the second and last time in mid-January 1834 to hear and adopt reports on the ten-hour day and a system of adequate wages. What evidence there is indicates that Manayunk's operatives, who had inititially turned out at the end of August, kept at least three of the six large cotton mills shut down through the first week of November.[40] However, we do not know precisely when or with what results the first strike ended. It is known that in the trough of the recession in early March 1834 mill owners on the canal simultaneously ordered wage rates reduced by 25 percent. It is unlikely that they laid a 25 percent reduction on top of one of 20 percent that had precipitated the strike in late August. In any case mill workers, now supported by Manayunk's wage earners and union leaders, organized swiftly, indicating that the strike committees of the fall may not have entirely disbanded. The three-month battle between labor and capital that now began revealed that factory workers, as well as mill owners, had resolved to toughen the tactics used the preceding fall.[41]

When Borie announced the 25 percent wage cut in March, his Schuylkill Factory operatives immediately gathered to plan their strategy. On a Saturday afternoon in mid-March, the first public meeting of "The Hands Employed in the Schuylkill Factory" was held behind the Presbyterian meeting house "to adopt measures to withstand the attempt made" by Borie to reduce their wages. James Hudson, a convert of the Presbyterian revival the year before, was called to the chair by his fellow factory operatives. William Gilmore, one of the leaders of the strike of late 1833 and president of the TUP, served as secretary of this first meeting. Three men of varying backgrounds addressed the crowd. One was William F. Small, the lawyer, and political radical who had been instrumental in the textile workers' opposition the previous fall. William T. Young, a Manayunk cordwainer, also spoke. A former member of Roxborough's Workingmen's Party of 1829 and 1830 and anti-Jackson party of 1832, he represented the established tradesmen of the township. Like James Hudson, he had participated in the evangelical revivals of 1833–34, which evidently did not function to counteract labor discontent and independent action. His involvement in the strike signaled the alignment of the politically active, conservative-minded tradesmen of the community with the cause of Manayunk's industrial workers. Sharing the rostrum with Young and Small was Samuel Ogden, the mill operative and Jackson man who had labored in textile factories in England and America for most of his eighty years. He well understood the grievances of the mill workers, for his four daughters were currently employed by Ripka as powerloom weavers.[42]

These men wrote the resolutions of the assembled workers and presented the rationale for their strike to their fellow citizens. They used less militant language than their predecessors, but they returned to the same inequities of the present system and the rights of wage laborers. "We will be in a far worse situation than the paupers of our township" if the wage reduction is not resisted, the strikers announced, since the wages they received before the 25 percent cut were already "barely sufficient to procure the common necessaries of life." The mill workers pronounced themselves as the most valued members of the community. "We are (without vanity on our part)," they asserted, "acknowledged to be the producers of all the wealth of society." Common justice required that they get more than a "scanty pittance" in return for their labor. "We are willing to do our duty as operatives," the workers proclaimed, "so long as our employers are willing to pay unto us prices by which we can obtain a living." While granting their position as factory operatives, they denied the prerogative of capital to set the value of their labor. They asserted, in the

The Mill Strikes and the New Ideology of Class

tradition of England's trade unionists and factory workers, their right to resist violation, in the form of attempts to reduce their wages, of "their only real property": "Our wages are our rights over which no one but ourselves have any just authority.... We will maintain and guard them from any and every encroachment whatever."[43]

As operatives from the Manayunk mill formed picket lines around the canal factories, the mill owners took steps to undermine the strike. One offered to reduce wages by only 15 percent in an unsuccessful attempt to induce his workers back to their machines. Borie simply sought replacements as he had done during the mule spinners' strike of 1828. Placing an advertisement in Philadelphia's *United States Gazette* on April 25, he announced that ten or fifteen families could find employment at his mill "where liberal wages would be given, and payment made every two weeks."[44]

The workers' committee countered Borie's attempts to break the strike. They printed and distributed a handbill warning families not to come to Manayunk to take the jobs of those who were standing out against Borie's reduction. Twenty-five to thirty families and a "large number" of single men and women in Borie's employ, the notice stated, were "willing to work at Liberal Wages—such wages as will procure them the necessaries of life, which they [Borie and Co.] are not willing to give them." The strike committee also took to the streets to discourage strike breakers. In late April, spinners James Hudson, chairman of the first strike meeting, and William McIntyre confronted a spinner who had offered to work two mules at reduced wages. They threatened him, according to one witness, then threw him to the ground before a constable could intervene.[45]

As the workers' standout wore into the second month, Borie's determination to break the strike hardened. At the end of April he enlisted Francis Barat to work as a spy to inform him of the names and actions of the members of the strike committee. Barat worked as a machinist in the Schuylkill Factory and his loyalty to the company's owners, Borie and Keating, had deep roots. Barat, like Borie, was a French Catholic and a refugee from the French Revolution who had arrived in Philadelphia just after the turn of the century. He had operated a small machine shop in the port city and moved to Manayunk in 1829, at Keating's request, to take charge of the Schuylkill Factory machine shop. He led the list of those whom Keating organized into the first Catholic parish in Roxborough and was completely devoted to him. While he was not bound so closely to Borie, Barat eagerly defended his employer against his fellow mill workers, asserting Borie's "right to procure hands" to break the strike. "You know I am a tried friend," he wrote to Borie in the midst of the strike, "and will be always ready to support you."[46]

Borie did indeed ask Barat to identify the strike leaders who penned the handbill warning families not to come to Manayunk seeking jobs. Barat's task was facilitated by the fact that members of his Catholic congregation were leaders on the standing committee. Barat knew enough about some of his fellow laboring men and parishioners to reveal something about the hierarchy of the committee. He informed Borie that two shoemakers, William Gilmore of Philadelphia and William T. Young of Manayunk had penned the broadside while Manayunk tavernkeeper Daniel Hughes and dresser Michael Gallagher may have contributed to its composition. The spinners on the committee he considered "too ignorant" to formulate the statement on the handbill, although they were "just as guilty" as the others.[47]

Barat's information probably influenced Borie's decision to bring in an armed "peace officer" to protect the strike breakers and to solicit the clergymen of the various churches to urge their congregations back to work at the reduced wages. By early May, however, support for the mill operatives had grown among the working class in Manayunk, and a larger and bolder strike committee publicly condemned Borie's tactics. When the "working people formerly in the employ of the Schuylkill Factory" met on May 7, they were on the offensive. Two female operatives, Martha Dyson and Ann Matsen, joined John Ferral to draft resolutions that condemned the illegal use of magistrates as guards to protect strike breakers. As long as the standing committee could procure supplies of food for the strikers, they expected the mill workers to remain on strike. The meeting's denunciation "of a few of our class" for advocating a return to work at the reduced wages suggests that the ranks of the wage laborers in Manayunk had coalesced in opposition to Borie. The fact that the strike had endured for two months also indicates that the community was providing "a sufficiency of aid" for those from the mills.[48]

The proceedings of the meeting published in the *Pennsylvanian* contained an unusual proposal to solve the impasse between laborers and capitalists, which revealed the mill operatives' recognition of the dilemma of the competitive system. The mill employees offered to pledge 25 percent of their wages, the amount of the proposed reduction, to fund the procurement of new machinery for the mill. They believed that Borie was not "able to meet his competitors in the cotton market" because of the inferior quality of his machinery, and therefore "oppressed [his] workmen by reducing the wages." If Borie would acknowledge his inability to "renew his machinery," the wage earners were willing to take over ownership of the machines (their investment to be secured by a mortgage) until the textile capitalists could buy them back.

The Mill Strikes and the New Ideology of Class

The meaning of the proposed compromise is perplexing. Did the mill operatives see Borie only as a victim of the crisis-prone market, like themselves? They may have understood the long-range impact of new machinery on the price of their labor and, at the same time, known the disastrous consequences for them if Borie failed. By 1831, John Doherty himself had acknowledged the need for the cooperative ownership of machinery in order to distribute "its benefits" more fairly. However, the tone of the workers' proposal suggests that they were not presenting an ingenious plan to take control of the machines that oppressed them. While the workers were not denouncing the system of industrial production, they were nevertheless carrying out a radical strategy in refusing to let the owners of the machinery dictate wages. In subsequent resolutions, they reiterated their rights as wage earners and condemned the system of "toiling from dawn of day till the shade of night, for that which will not procure a sufficiency to support nature."

After May 1834, there is no further mention of the mill workers' scheme. Within a week, the Schuylkill Factory hands met again in a less conciliatory mood, angered by the most recent tactics of the mill owners. The acting committee included Martha Dyson, Samuel Ogden, his daughter and powerloom operator Sally, John Ferral, and William Gilmore. They drafted resolutions denouncing Borie's continued use of "magisterial aid" to break up assemblies of strikers. Borie had also found an important ally in Manayunk's clergy. Some, if not all, of the town's ministers intervened directly in the conflict on the side of capital. In their resolution, the Schuylkill Factory hands condemned the pastors of Manayunk's churches who "forced some to go to work at the reduced prices." While it is unknown precisely what the pastors were doing to coerce workers to return to the mills, the accusation indicates that the tension of the conflict permeated the recently revived congregations. The convened strikers also publicly thanked one of Manayunk's own magistrates who had protected them from the "unlawful measures of [their] employers." The Schuylkill Factory employees were as determined "as when we first struck, to withstand each and every attempt to oppress us."[49] Before the end of May, just as the recession bottomed out, the mill workers returned to the mills, seemingly victorious. While some had obtained jobs elsewhere, those who went back to the mills apparently did so with a 5 percent increase in their old wage rates.[50]

The settlement of the Schuylkill Factory strike in May 1834 ended a nine-month struggle between mill owners and mill workers. The importance of the strikes of 1833–34, however, transcended their impact upon the relationship of labor and capital within the mill. They had taught the community about the antagonisms that characterized the factory system. The factory operatives' cause had also given life to the insurgent labor union movement in Philadelphia. The nine-month battle in the mill town propelled radicals Gilmore and Ferral into positions of leadership in the Philadelphia and then the national trade union movement.

The political career of Ferral, the Irish-born handloom weaver, was even more meteoric than that of Gilmore. After Ferral had helped orchestrate the Manayunk strikes in the fall and spring of 1833–34, he traveled to New York that summer as one of seven Philadelphia delegates to the convention of the National Trades Union, the first nationwide organization of wage earners. Ferral's name was recorded on the list of delegates as representing the Manayunk and Blockley manufacturers. He brought to the attention of his fellow representatives the condition of female laborers in manufacturing establishments. While Ferral was a handloom weaver who had never worked in the mills, his long involvement with the Manayunk operatives' cause had familiarized him with the exploitative system of labor—"the 14 to 16 hours per day in watching a few threads, or the moving of a shuttle . . ., the impure air." The mill operatives' strike had also educated him about the power of the industrial capitalist. "At present," he told the convention, "the capitalist has all the protection, all the privileges, and all the power, [he] can reduce the measure of their pittance—add to their hours of labour—or turn them out to beggary!"[51] Like Doherty, Ferral viewed society as divided by class and saw the trade union and independent working-class action as the means of reform. In the following year, Ferral was unanimously elected president of the National Trades Union.

The Manayunk mill operatives' movement fed the class-conscious content of the trade union movement and left an indelible mark on the social relationships of Roxborough, exposing and nourishing an enduring conflict that went beyond the battle over wage cuts between recession-plagued employers and employees. Factory women and children and male operatives had picketed, demonstrated, and debated in the streets along the canal against the capitalist's claim to determine their wages. Manayunk mule spinners, shopkeepers, and wage-earning tradesmen—Irish Catholics and Protestant revivalists alike—participated in and led the public strike meetings that proclaimed laboring people to be the producers of the wealth of society and condemned the textile capitalists for denying them their just portion of that wealth. Philadelphia political and trade union radicals, steeped in British oppositional tradition, assisted the Working People of Manayunk. They described owners and operatives as separate and hostile

The Mill Strikes and the New Ideology of Class

forces, articulating an anticapitalist ideology and working-class critique of the situation in the mill town.

In 1837 Charles Hagner, writing as "Plain Truth," recapitulated what he had "seen and observed" in his community for the past decade. "Combinations of the employers," as Hagner referred to the mill owners, had acted and spoken aggressively to protect and advance their prerogatives and interests as early industrialists. He lamented the disturbing effects of the " 'strikes' and 'turnouts' that have so frequently occurred in this village." Hagner pinpointed the significance of the long season of mill-worker protest. "All have a direct tendency," he wrote of the strikes, "to array one portion of the community against the other, the poor against the rich and the rich against the poor." He voiced the same alarm expressed by Englishman Henry Tufnell in 1834 in his analysis of the effects of trade unionism. "A complete separation seems to have taken place between masters and men. . . . Each party looks upon the other as an enemy."[52] A new sentiment, one of class consciousness, had begun to emerge in early industrial society on both sides of the Atlantic as English cotton spinners, handloom weavers, and trade unionists carried the new working-class ideology from English to American factories.

The strikes of 1833 and 1834 had not only widened the gap between employer and employee but had allowed the community's tradesmen and small proprietors, as well as factory workers, to make the link between the issues and struggles in the mills, the afflictions of the competitive industrial system, and political power. The spring of 1834 witnessed the high point of working-class unity in early industrial Roxborough, as wage earners turned the local Jacksonian Party into a radical anticapitalist coalition to attack the bank and the industrial capitalists of the community.

VIII

THE POLITICS OF CLASS CONFLICT IN EARLY INDUSTRIAL SOCIETY

The issues over which the mill workers and owners of Manayunk struggled in 1833 and 1834 became topics of concern, too, for farmers, craftsmen, merchant millers, and the small proprietors of greater Roxborough. The strikes had clarified the depth of the antagonisms that divided labor and capital. Spokemen for both sides had laid bare in the local press the fundamental views that set them against each other. Women and child operatives had formed picket lines around the mills and marched through the streets. Mill workers from around the region rallied behind the striking Manayunk operatives and together they molded the first broad-based industrial trade union in the country. Increasingly, the mills of Manayunk appeared to Roxborough's old elite and tradesmen to challenge "community notions of governance and social hierarchy."[1] In the months that followed the strikes of 1833–34, those concerns were deepened as the conflict within the mill gates permeated party politics and the mill owners of Manayunk took the offensive.

On May 27, 1834, the *Pennsylvanian* reported that "Manayunk has been completely revolutionized by the Panic measures of the opposition party." The depression, precipitated in August 1833 when Nicholas Biddle curtailed credits from the United States Bank, peaked in May. A few days earlier, the Jacksonian press had observed "the largest public meeting ever held in Manayunk on any occasion." Over one-quarter of Roxborough's male citizens crowded into the yard of the Fountain Inn and put their names to resolutions that supported President Jackson and condemned the mill owners' actions in the strike. Two hundred and fifty-two laborers, spinners, factory operatives, carpenters, carters, coopers, shopkeepers, farmers, and various tradesmen inscribed their names as Democratic Republicans who supported the "patriotic chief Magistrate" in the political warfare being waged by the bank and other "aristocratic" and "monied" institutions.[2] The former strike committees were well represented at the meeting. William Gilmore served as secretary along with Manayunk shopkeeper George W. Davis.

The conflict between labor and capital in the mill seemed to have refashioned prestrike political alignments. Manayunk's Jacksonian Democratic Republican Party cut across religious divisions and included small proprietors and wage earners alike. Skilled Protestant craftsmen, who in prestrike days would likely have joined the anti-Jacksonian party under an evangelical banner, signed their names alongside Catholic laborers and shopkeepers. The elected leadership of the meeting on the twenty-seventh reflects this diverse makeup. Methodist Amos Phillips, an operative by trade and the constable who had defended the strikers against the "unlawful measures" of Borie and other mill owners in April, was elected president of the meeting. Catholic laborer William Welsh joined longtime innkeeper Michael Snyder and drug-mill owner Charles Hagner as vice-presidents. A physician and a cooper were appointed to draft the resolutions. Significantly, the only groups that were not represented at the Jacksonian assembly were the textile capitalists and Roxborough's Wissahickon merchant millers and farmers.

It is clear from the *Pennsylvanian* article that after nine months of economic recession, political party tactics as much as the issue of the bank had brought the Jacksonians and Whigs to blows. Roxborough's Jacksonians emphasized that they reluctantly entered the system of party warfare that their adversaries had introduced into the town. Their Whig opponents had thrown down the gauntlet, however, and the days of peaceable and quiet voting were now gone, they announced. They charged the Whigs with the base tactics of organizing "a political society," adjourning public meetings, and using proscriptive measures to obtain votes for the bank. The Jacksonians stated that they had no choice but "to show our friends abroad that Manayunk is not . . . on the side of the Bank."

Manayunk's Jacksonians squarely placed the textile mill owners in league with the bank aristocracy and its retinue of privileged monopolies. The aggressiveness of the textile capitalists in trying to affect the outcome of the elections demonstrated how the bank war had evolved into an issue of conflict between labor and capital. At a meeting on May 23, Manayunk Democratic Republicans denounced one of the mill owners who had discharged six "American citizens" for stating their opposition to the bank. This "high crime" against those who exercised their "natural and political rights," the Jacksonians charged, proved that the "principle of modern whiggism" was the same as the "federal, tory principle" that lay behind the Alien and Sedition Acts of 1798. Two weeks before the Jacksonians gathered to present these charges, the Schuylkill Factory strike committee had accused Borie of the identical "act of high-handed tyranny." The principles of "the industrious working-people," the Jacksonians announced indignantly were not to be affected, like certain sectors of the capitalist class, "by the price of cotton, tobacco, and bank stock."[3]

Economic coercion at the polls was understood by Jacksonians throughout the Philadelphia region to be a major tactic of the Whigs. During the fall campaign of 1832, the *Pennsylvanian* posed the question for the wage earners: Do the "task masters ... not require you to vote as they please, or quit their employment?" When Ripka's employees struck in August 1833, they charged that he had "rob[bed] the operative of his rights and privileges as citizen, as well as the common necessaries of life." That fall John Ferral accused Manayunk mill owners and "bankites" of using despicable tactics in their campaign to uproot Jackson. He accused "hirelings" of the bank of trying to convince the workingmen of the village to vote against Jackson in the 1832 election and reproached the mill owners—"the *spinning jennie nabobs*"—for attending the polls "to coerce if possible the freedom of the election."[4] On April 5, 1834, around the same time that the Schuylkill Factory strike committee and Manayunk Jacksonians accused Borie of political harassment, the *Pennsylvanian* printed the names of a committee whose purpose was to "ascertain the names and residences of all those who have been thrown out of employ, for honestly entertaining an opinion in opposition to their employers."

When the Manayunk Jacksonians reconvened with their Roxborough allies on the last day of May, in what was claimed once again as "the Largest Public Meeting ever held" in the township, they denounced the coercion of the industrial mill owners. Samuel Ogden, who had spent a long career in the mills, spoke before the meeting about Joseph Ripka's firing of employees for their political views. Relating his personal experience of political awakening, Ogden told the hundreds of tradesmen, laborers, storekeepers, and farmers from the township that he had been a naturalized citizen for twenty years but up to this point had never voted or "meddled with politics." Jackson's battle with the bank, however, had mobilized him, and he could no longer refrain from "expressing his opinions on the subject" and supporting the president. For his actions, Ogden informed the crowd, Ripka had discharged his four daughters who worked as powerloom weavers.[5]

The spring of 1834 proved to be a period of political enlightenment for Roxborough's working class. The wage earners of Manayunk turned to party politics for the same reason that the mill workers had aligned in the Trades Union of Pennsylvania the previous fall. Like the trade union, the political party was perceived as a class-specific institution to combat encroachments upon the economic and political rights of workers. Not only factory operatives like Samuel Ogden but wage earners of every occupation became convinced that the mill owners were making a mockery of the democratic system. Borie and Ripka were viewed as nothing less than aristocratic despots. When a quarter of Roxborough's men gathered en masse to endorse the Jacksonian platform, they were using the party much as they had used the trade union—as a vehicle to counter the political and economic tyranny of the industrial capitalist class. In a meeting in early June 1834, they questioned the very efficacy of the industrial system. The rupture between employer and employee was depicted as a laborers' struggle to protect sacred liberties against capitalists and despots who were one and the same. If the "effect produced," the Jacksonians stated,

> by the large manufacturing establishments is to enslave the minds of the citizens employed in them, and if they are, . . . through fear or favour, to be led in mass public meetings, jubilees, &c, contrary to their own convictions, . . . and if for the free exercise of the invaluable right of suffrage; . . . they are to be proscribed and turned out of employment, it is high time for the American people "to count their cost" and inquire whether such establishments are the best preservatives of our free institutions.[6]

De Tocqueville had foreseen the dangerous consequences of a "manufacturing aristocracy." Now "a permanent inequality of conditions and aristocracy" indeed seemed to be entering the world of early industrial Roxborough through the mill gate.[7]

The mill owners on the canal took the verbal attacks by Roxborough's Jacksonians as a serious threat. Their participation in the fall of 1834 in an unprecedented political coalition with Roxborough's merchant millers and land-rich farmers was a tactical countermeasure to the challenge of the working class. In September and October, the Democratic Whigs of Manayunk organized to endorse candidates and nominate delegates for the local and congressional campaigns. Mill owners Joseph Ripka and John Rush served on various committees for the Whig Party as did Wissahickon millers Nicholas Rittenhouse, Jacob Rittenhouse, and Jonathan Robeson. Of the fifty-eight citizens who turned out for one of the Whig meetings in September, only one laborer and three mill workers can be identified as in attendance.[8]

The campaign strategy of the Whigs in the race of 1834 was to divide the large working-class Jacksonian bloc. Manayunk's local Whig Party tried to exploit Irish sectarian differences and anti-Catholicism in order to draw Protestant immigrants to their fold in the fall election. In order to break the Democratic grip on the county, Whig leaders ran James Gowan, an Irish Protestant tradesman, in the congressional race against Southwark's Jacksonian Joel B. Sutherland.[9] The tactic did not succeed in Manayunk, however, because the town's tightly constituted, nonsectarian Friends of Ireland, founded in May 1833, headed off the potential split. American, English, French, as well as Irish, had formed this association to support Ireland's cause and "the cause of Civil and religious liberty in general."[10] James Gowan was a popular supporter and promoter of the Friends of Ireland movement in America. As a Whig candidate, however, he also backed the rechartering of the United States Bank. At the enormous Jacksonian meeting at the end of May, the guest speaker, Catholic judge Joseph M. Doran of Southwark, attacked Gowan for supporting the bank. The following July, as reported in the *Germantown Telegraph* on the ninth, the Irishmen of Manayunk met as a partisan group of Whigs for the purposes of "expressing their disapprobation" at Doran's disparaging remarks against one of their countrymen. The ad hoc committee that formed to assure Gowan of their support was indeed composed of the nucleus of the township's activists.

At this point, John Ferral skillfully intervened to preserve Irish backing for the Jacksonian candidate and the working-class party of opposition. In early August he addressed the large meeting of the Irishmen of Manayunk. While reprimanding Doran for insulting a fellow Irishman, he tactfully chastised Gowan for lending his aid to "the failing edifice of Bankism." Ferral urged the Friends of Ireland to pledge themselves to Jackson in order to "clear these United States of the hireling slaves of British aristocracy." Two months later Roxborough gave Sutherland, who won the election, a sound majority over Gowan, as did the other working-class townships in the county. Roxborough's voters also elected Jacksonians to local offices.[11]

By the end of the fall campaign of 1834, the battles between employers and employees in the mill yards and at the polls had convinced Roxborough evangelical leaders that their community had become dangerously polarized. In the months that followed, Baptist and Presbyterian patriarchs and their wives launched a local temperance and moral reform campaign against what they considered the divisive forces of early industrial society.

To understand this movement, it is necessary to go back to the summer of 1833 and follow developments in the churches after the revivals of the previous winter. During that time, Baptist tradesmen and a Presbyterian mill owner had helped conduct the working-class awakening in an attempt to impose the moral sanctions of Protestant religion—self-discipline and restraint—upon the industrial laboring class. Within months of the end of the revivals, they learned that evangelical conversion alone could not smother economic and ideological conflicts. In August 1833, the Baptist patriarchy began to express concern for the fact that several brothers and sisters were "habitually absent[ing] themselves from the communion of the church."[12] In trying to explain Nichols's pastorate and the period of reaction that immediately followed the revivals, Reverend James Willmouth, the Baptist pastor of Roxborough in the 1880s, suggested "there must have been somewhat too much of mere excitement, and that persons were too hastily judged to be converts and baptized."[13] Perhaps more to the point, the beginning of this period of declension coincided precisely with the onset of the mill strikes.

Many of those mill workers and wage laborers who participated in the "extensive work of grace" in the winter of 1832–33 probably participated also in the strike movement. As they did, their enthusiasm for Baptist communion or precepts may have waned. No lists of the rank and file members of the strike committees of 1833 and 1834 exist to prove this point. However, of the handful of strike participants and leaders who can be identified, six were among the revival converts. Indeed, Presbyterian converts Ann Matson, spinner James Hudson, machinemaker William Singer and Baptist converts William Young and Jacob Kuhn, both shoemakers, resembled the labor leaders among the Primitive Methodists of England, "whose political consciousness and activity began with or shortly after conversion."[14] For workers like Ann Matson, who perhaps labored in Darrach's mill before joining his church, the act of shared conversion became a springboard for the unity and zeal necessary to challenge their employers.

As the Working People of Manayunk organized the initial strike committees and determination rose to resist the wage reductions, the Baptist church began to crack down on those who were abstaining from communion. On August 27 the Baptist meeting sent a discipline committee to investigate those who had missed three regular communion lessons. On September 14 they formed a standing committee of inquiry, consisting of established tradesmen and anti-Jacksonian party members Charles Levering, William Simpson, and William Rawly, to visit delinquent members. At the same meeting, the leadership passed a resolution requiring those found guilty of disorderly conduct to make an open confession to the church before being restored to fellowship.[15]

The warning and exclusion of digressing members, a practice as old as the congregation itself, became prevalent in the months of mounting strike activity. Between the spring and fall of 1834, a period bracketed by working-class victories in the mill and at the polls, the Baptist church excluded five members, the most in any single year of its history. Among the five was Jacob Kuhn, who as we have seen had been converted during the revivals of January 1833 and had joined the first strike committee in early August of that year. In April 1834 the Baptist meeting excommunicated him "for not believing himself converted." William Singer experienced the similar wrath of the evangelical Presbyterian church. Singer had been converted during the Presbyterian revival in April 1833. During the spring of 1834, he served on the standing committee of the striking mill workers. Following their victory, he participated in the massive Jacksonian organizational meeting. Not surprisingly, he was subsequently suspended from the church founded by James Darrach.[16]

So ridden was the Baptist church with dissension at the end of 1834 that Nichols offered to resign, hoping that a different pastor could bring the church back "together as one solid compact." With the Baptist leadership opposing the resignation of their gifted pastor, Nichols stayed and helplessly watched the decline of his evangelical congregation over the following year. In his annual report to the Baptist Association in October, Nichols described the church as being "in a dull and languid state, owing . . . to neglect of duty and want of faith in promises." The number of exclusions peaked at seven the next year, the highest in the congregation's forty-five-year history.[17]

It was during these months of "general coldness" in the Baptist fellowship that Baptists and Presbyterians turned to moral reform as a means to bring order and unity to their congregation and community. As Nichols reported, they entered into "the benevolent objects of the day such as the Missionary Tract, Sabbath School, Education and Temperance Societies."[18] On March 19, 1835, the Baptist meeting resolved that "the habitual use of ardent spirits" would disqualify any member from communion, and as a church took the pledge of abstinence. Of the record number of seven members who were excluded that year, four were charged with being intoxicated.[19]

In September 1834, six months before the Baptist meeting resolved to promote the cause of temperance, the Flat Rock Temperance Society was established by six persons who signed an oath of total abstinence. By May 1835 the society had held thirteen public meetings, distributed 1,500 temperance publications in the mill town, and recruited 183 members who took the abstinence pledge.[20]

The first semiannual report of the society, published in the *Germantown Telegraph* on May 20, suggests the particular function of temperance ideology in an industrial factory town. Protestant employers aimed particularly to reform the mill operative, who not only imbibed quantities of spirits but also the anticapitalist critique explicated by John Ferral and other working-class radicals. They recruited in a neighborhood "proverbial for rum-drinking, frolicking and fighting," as well as labor militancy and radical Jacksonian politics. Manayunk's Protestant employers tactfully portrayed the society as a group of self-directed workingmen. We are, they insisted, "emphatically an operative or workingman's society," having started "without the aid or countenance of the talented and influential members of the community." The temperance association members, however, felt obliged to add, "justice requires us to state that ... from some few individuals (who cannot strictly be called workingmen) ... the society had received pecuniary aid and efficient support." One such individual was one of the Presbyterian, mill-owning Darrach brothers. He was in attendance at the meeting and helped to compose the report, which, in essence, directed the employers on the canal to clean up their workshops. None benefited more from temperance reform, the society pointed out,

> than proprietors of manufacturing establishments; everyone knows that the direct and unavoidable effect of ardent spirits is to [render] the person who indulges in them, totally unfit to superintend the movement of complicated machinery; ... we therefore feel surprise that all persons engaged in carrying on manufactories, do not feel it a duty ... to promote the cause of temperance.[21]

While intemperance "to a fearful extent" was still prevalent in Manayunk in 1835, so was trade union agitation and political radicalism. It was no coincidence that Manayunk's temperance society was founded the same month that Manayunk textile capitalists and Wissahickon millers organized into the Whigs of Manayunk to oppose the Jacksonians. But the town's proprietors, particularly the textile mill owners, had little time to practice the preachings of the Flat Rock Temperance Society "to promote the cause of temperance." Within a month of the publication of their report, Manayunk's laboring class organized and confronted their employers with their own demand for reform of the factory system.

In June 1835, Manayunk became a focal point in labor's push for a new system of hours. On June 15, mechanics, mill operatives, and laborers of the mill town met in the open square in front of the Manayunk Academy to "take into consideration the utility of the 10 hour system of labor." Tradesmen, railroad laborers, mill operatives, machinemakers, and weavers enthusiastically canvassed all employers to determine whether they would accept the system. Responding to pressure already applied by their own workers, most of Manayunk's employers assured the committee that they were willing to give the reduced hours. Their names were read before the assembled crowd of workers to the reception of hearty cheers. The names of "two or three gentlemen," who had ignored the requests of the working people, were greeted with three groans.[22]

The employers, mill owners among them, who refused the demands of their hands for a reduction in working hours met with more than verbal censure. Within days of the meeting of Manayunk's "useful classes," the operatives in the canal factories went on strike. They joined laborers and wage-earning tradesmen throughout the county in the nation's first general strike for the reduction of the hours of labor. The idea of a general strike was a radical one, introduced by English Owenite William Benbow in 1831 and distributed widely in pamphlets the following year. Apparently inspired by John Ferral's reprinting of the "Boston Circular," which assailed the long hours of Boston's workers, twenty thousand wage earners in Philadelphia walked off their jobs in early June demanding the ten-hour day.[23] Ferral, who had witnessed the aggressiveness of female textile operatives in Manayunk during the strikes of 1833–34, was again impressed with their militancy in the ten-hour strikes. In his typically evocative style, he wrote Seth Luther in Boston that "even oppressed females, and children employed in cotton mills in Manayunk (those brutalizing emporiums of human misery) have caught the spark of freedom's fire and are now on strike for the hours." Within a week of walking out of the mills, operatives received assurances from their employers that their "days service shall close at a somewhat earlier hour."[24]

Whether or not the "blood-sucking aristocracy," as Ferral called Philadelphia's employers, were "terror stricken" by the ardor of the working people in Manayunk and Philadelphia, they felt compelled to organize an attack in the newspapers against the general strike and ten-hour movement. The conservative Philadelphia *U.S. Gazette* published an editorial in June 1835 entitled "Labour and Capital." It argued that workingmen, in the long run, would be hurt by such protests. Laborers needed capitalists, the editor counseled, because they provided everything for the laborers, including food and clothing. It was a "folly for workmen," the editor instructed the city's operatives and laborers, "to look with jealous eye upon 'their best friend.' "[25] On June 24, the more sympathetic *Germantown Telegraph* anxiously voiced its concern that the extra hours of leisure might be "applied to useless and unworthy purposes" and thus would precipitate "an equal unanimity in setting [the ten-hour system] down as pernicious and injurious" to laborers and capitalists alike. The sharp-tongued

Ferral had only this scathing advice for those who questioned whether the workingman or -woman would benefit from this reform:

> Say, then, ye would be philanthropists and temperance drivellers what can you expect your fellow man will do in this case? ... fly to the poisonous intoxicating draught? Know ye panderers to inequality and tyranny ... ye plunderers of the *poor* that the whole of the errors and crimes of society are chargeable on your heads.[26]

Most trades were successful in attaining the ten-hour work day through the general strike of June 1835. This in turn propelled thousands into the year-and-a-half old Trades Union of the City and County of Philadelphia. The general strike also brought to the attention of radical leaders other than Ferral the plight and militancy of the female and child operatives of the Manayunk mills. In his presidential address at the celebration of the mechanics' TUCCP on July 4, the less strident William English condemned the cotton mills of the county as "modern hells" and called for the formation of a Female Trades Union.[27]

The working class's success in the ten-hour movement and the growing strength of the TUCCP fueled sentiment for an independent workingmen's party. The seeds for the third party movement were planted during the 1834–35 state legislative session when the leadership of the Democratic Party split. During the winter, the party's legislators divided over the issue of the public school law and Jacksonian governor George Wolf's decision to seek a third term in office. Disaffected Democrats, the workingmen of Philadelphia among them, formed an anti-Wolf faction for the gubernatorial race of the fall of 1835. Considering the laboring classes to be a decisive force, Wolf's party opponent, Henry A. Muhlenberg, made the decision to court the vote of the workingmen's political societies. When the countywide Society of Mechanics and Workingmen addressed a questionnaire to the two candidates in March 1835, Muhlenberg tactfully endorsed their entire platform. With the gubernatorial candidate declaring his adherence to the goals of public education, militia law reform, penitentiary reform, and most significantly, the six-to-six work day and opposition to the bank, the workingmen rallied behind him in what was, in fact if not in name, a born-again Workingmen's Party.[28]

During the general strike in June, English-born tradesman Thomas Brothers began publishing *The Radical Reformer and Workingman's Advocate*, which became a forum to advance the political goals of Philadelphia's working-class movement. On June 13, during the week-long general strike, Brothers attacked the orthodox Jacksonian Democratic Party for abandoning the interests of the working class and supporting banks and chartered monopolies. Two weeks later, in a diatribe against the usurping aristocracy in Harrisburg, Brothers announced that workingmen "have the power to elect an assembly in the fall to undo all that has been done by the present one."

Indeed, in attending the meeting of Roxborough's workingmen's Democratic faction during the fall campaign, one would have been transported back to the prestrike days of the moderate Workingmen's Party of 1828. The old reformer Charles Hagner, who had run for the state senate in 1829 on the Workingmen's Party ticket and who had been a leader of the Jacksonian party in 1834, now chaired a meeting of the Democratic Citizens of Roxborough in August which pledged support to a general system of education, complete reform of the militia system, and opposition to the chartering of any new banks or stock-jobbing companies.[29]

Organizing in the same months was Roxborough's local Whig Party, which met in September to pledge their support to probank candidate Joseph Ritner and to elect delegates to the county convention. The Whigs of 1835, like those of 1834, consisted of a coalition of early industrial capitalists, Wissahickon millers and commercial farmers, and established tradesmen. Mill owners such as James Darrach and James C. Kempton, perceiving the need to prevent a working-class political victory, for the first time took an active role in the township's party politics, as Joseph Ripka and John Rush had done the previous year.[30]

In October both the Muhlenberg and Wolf Democratic tickets were defeated in Roxborough and in the rest of the county as well. Whig candidate Joseph Ritner received a large plurality in all districts except the Jacksonian stronghold of Southwark. He garnered a little more than one-third of the votes in Roxborough (where just 42 percent of the taxpayers went to the polls). But the Democratic majority split evenly between Wolf and Muhlenberg, suffering their first defeat in the industrial township since 1832 and carrying the working-class program down with them.[31]

There are no extant lists of Roxborough's "Democratic Citizens" of 1835 that would enable us to sort out the factional split in the township's Democratic Party. Conspicuously missing, however, from the platform of the August meeting was a commitment to the six-to-six work day, an issue that strongly appealed to Manayunk's factory workers. The short-time issue was singularly important to contemporary Lancashire cotton workers and their political activity. If Hagner and other Democratic leaders had left it off the pro-Muhlenberg agenda, they may have alienated the support of a good portion of Roxborough's wage earners. In any case, only 259 of the township's Democrats turned out to vote for either Wolf or Muhlenberg. It is clear that a coalition of wage earners did not have the power to split from the Democratic Republican Party and run their platform successfully against the Whigs and regular Jacksonians. The Whig victory in the fall of 1835 proved to be an important turning point in the political battle between labor and capital in both the state and Manayunk.[32]

Following the electoral victory in the fall of 1835, the Whig-dominated legislature moved to squelch the working-class presence in the electorate. Two bills enacted in the following spring were designed to cut the working-class base out of the opposition party. One reapportioned the political districts. From the point of view of Manayunk's working-class voters, the Apportionment Act was "unequal, unjust and unconstitutional." It gerrymandered the assembly districts to give the "patrician and aristocratic city of Philadelphia" seven representatives for 18,000 taxables and the county and its working-class neighborhoods of 31,000 taxpayers eight representatives. In the same session, the Whig assembly also passed the Registry Law, which required annual registration of voters. The Democratic Republicans recognized this as a method to deprive the workingman of his vote, since the wage earner would be reluctant to leave work and lose essential earnings to register.[33]

The anti-working-class sentiments that generated the aggressive action of the Whig legislators also underlay the bold offensive taken by the Whig mill owners of Manayunk to end the political challenge of the Jacksonians or anyone else who questioned their legitimacy. Evidence of the coercive power of the industrial capitalists surfaced in the assembly election of 1836. In August and September, Roxborough's Whig Party for the first time divided into hostile camps over the choice of delegates for the county convention. They split cleanly into competing capitalist factions: the industrial mill owners of Manayunk and the land owners and merchant millers of the Wissahickon. At a meeting on September 3, fifteen chagrined farmers and merchant millers of Roxborough gathered to condemn the political tyranny of the industrial employers on the canal. Chaired by Nicholas Rittenhouse and using language reminiscent of the strike committees and Jacksonian meetings, the committee of Roxborough's old elite charged that it was

> well understood that certain aristocratic mill-owners of Manayunk, under the disguise of Democratic Whigs, gave special attendance [at the August 27 meeting to elect delegates] for the avowed purpose of preventing their workmen from exercising the rights of freemen.[34]

One of those mill owners who they characterized as a "political demagogue" was John Rush. Nominated by the Manayunk Whigs as a candidate for the assembly district, Rush defeated the Roxborough Whigs' choice, Robert M. Carlisle, by seventy-five votes.

At the Roxborough Whig meeting of September 3, the flour millers and farmers passed resolutions disapproving of Rush's nomination and declaring that he was "a purse-proud autocrat and political demagogue." They claimed that he had been "unceasing" in his efforts to silence unfriendly workmen. The mill owner had insulted the citizens of Roxborough by frequently declaring that they were "too ignorant" to hold a valid opinion on the candidates running for office. Sounding more like Jacksonian workingmen than bourgeois Whigs, Roxborough's millers and farmers charged that Rush was incapable of representing the "interest of the workingmen in the councils of our state." If elected they feared he would "use every exertion to have chartered a moneyed institution in the village of Manayunk that will benefit none but the aristocrats in the speculation of stock." Regardless of issues, the Roxborough Whigs were horrified that the textile capitalists had used the same tactics against them as they had employed against the Jacksonians in the election of 1834. After resolving not to support Rush as the Whig candidate, they turned their wrath on the other textile mill owners and employers of Manayunk who had discharged the employees who had opposed Rush. Roxborough's alienated Whigs agreed not to hold any future meetings in the textile borough.[35]

Six days later, Manayunk's Whigs gathered to present an audacious statement of the prerogatives of the industrial capitalists. On September 9 the Whig friends of John Rush met to answer the Roxborough Whigs' charges. After denouncing the Roxborough Whigs for "sow[ing] dissension in the ranks of the party," (emphasizing they were "by no means as powerful as they are malicious"), the Manayunk Whigs denied that any mill owner had ever discharged a man "solely because he differed in political or religious opinion."[36] They portrayed the textile mill owner as an honest producer:

> We have yet to learn that the pursuit of an honest and honorable business, giving employ and support to hundreds, is to entitle a man to the name of aristocrat, and that more especially if by industry and economy he has been able to acquire some property as the reward of his toil.

While the *Germantown Telegraph* did not print the names of those at the meeting, there is little doubt that textile capitalists James Darrach, John Rush, James Kempton, and Joseph Ripka were in attendance. A resolution defending repression passed easily when it was offered from the floor:

> We consider the Mill owners of Manayunk, and every person employing men, to have a better right to discharge any man from their employment for differences of opinion on political questions than either the President of the United States or Governor of the State.[37]

Industrial capitalists paid their employees from their own pockets, while the government used the "People's money," the Manayunk Whigs charged. The workingman had an equal right, they pointed out, to refuse to work for anyone whose political sentiments they opposed. The Manayunk Whigs laid out the capitalist creed that located the mutual interest of worker and capitalist in the freedom of contract: "Self-interest is the foundation of all contracts both on the part of the employer and the workingman [and thus] we have a right to employ whom we choose or work for whom we choose." In dismissing the charges of the Roxborough farmers and merchant millers as an "attempt to array one class of the community against another," however, the industrial capitalists unwittingly exposed the deep divisions that necessitated their economic coercion.[38]

The "sentiments and reasoning" contained in the mill owners' proclamation of rights did not escape the township's working class any more than it had the old elite of Roxborough. An "uncommonly large meeting of Democratic citizens," held two weeks after the Manayunk Whigs had met, denounced the industrial capitalists' abuse of their powerful position. The requirement of a workman, the angry wage earners and tradesmen protested, "to think as his employer thinks—vote as he votes, forms no part of the 'contract' and none but slaves, will suffer themselves to be dictated on this subject by anyone."[39]

On the eve of the five-year depression, it appeared that de Tocqueville's fearful prediction of the ascendancy of "the manufacturing aristocracy" had become reality in industrial Roxborough. At the meeting on September 9, the "aristocratic" mill owners of Manayunk announced that the borough's Whigs would hold no more meetings on the Ridge Road outside of their stronghold, reminding the disaffected Whig farmers, tradesmen, and millers of Roxborough that "Manayunk is the metropolis of the township."[40] For Roxborough's industrial laboring class, the end of 1836 marked a year of inaction—the first in four—during a time when weavers and spinners in cotton factories around Philadelphia and in Norristown were striking over wage reductions and hours of labor.[41]

The mill owners of Manayunk had not, of course, achieved political domination over the workers and proprietors of Roxborough. They had closed ranks, however, and emerged in the perception of the old elite as a distinct and hostile group whose political and economic interests were inimical to the rest of the community. This marked the culmination of a complex but discernible process in which class interests underscored political behavior throughout Roxborough's social order.[42] Men from all trades who had pushed for broad social reforms in the preceding decade within the reformist Workingmen's Party aligned with the less moderate mill workers to use the local Jacksonian Party to oppose the political goals of the industrial mill owners. The old-stock tradesmen and shopkeepers and their wives, who had led the evangelical revival of 1832–33, launched the temperance movement in hopes of containing the social conflict that periodically erupted in the mills. Previously quiescent flour millers and farmers founded the local Whig Party in alliance with textile mill owners to oppose the working class's political movement. Finally, the alarming tactics of the textile mill owners in the 1836 election made clear to the old elite that these new entrepreneurs sought authority commensurate with their economic power.

The Panic of 1837 and the five-year depression that followed brought an end to the industrial conflicts that had galvanized the entire community. The depression began to slow textile production in the mills in the spring of 1837 and soon closed many of Manayunk's manufactories. Deteriorating economic conditions smothered labor opposition in the mill town and deadened the working-class trade union movement throughout Philadelphia for the next five years. Between 1837 and 1842, many inhabitants left Manayunk to seek employment elsewhere. As the "gloom settled" on the economic activities of the textile town, however, the divisions and alignments, and the new sensibilities of class, clearly marked the new industrial order of Roxborough and Manayunk.[43]

The history of the coming of the factory system and the social relations of industrial production to Philadelphia's textile districts is an important chapter in the story of American class development in the early republic. It was a chapter written largely by the immigrant textile workers as they acted to control the world of capitalist production and interacted with others to mold the institutions and relationships of a factory community. To have characterized Manayunk as "the Manchester of America" and its mills as "modern hells" was to make a vivid allusion to the worst manifestations of class exploitation and conflict in the unregulated, competitive world of the early industrial textile economy. Why the dynamics of class conflict distinguished the process of mechanization and factory development in Philadelphia, and not southern New England and elsewhere until a later date, is largely explained by the distinctive character of Philadelphia's labor force.

While New England's mill owners drew from a relatively scarce rural labor pool, the urban seaport's textile capitalists received an abundant supply of skilled weavers and spinners who had emigrated from Britain's stagnating textile districts. Between 1790 and the early 1830s, only periods of hostility between America and England stopped the flow of impoverished men, women, and children into Philadelphia's workhouse manufactories, handloom shops, outwork systems, and subsequently, into the cotton textile mills of Manayunk. By 1820 Irish and English male handloom weavers and mule spinners had turned Philadelphia into the center of fine yarn and cloth production. Thereafter, immigrant women and children, disembarking in growing numbers in the 1820s, filled the demand for powerloom weavers and machine attendants in Manayunk's mechanized cotton factories.

The particular skills and experience that the British textile laborer carried to Philadelphia had a profound effect on mechanization and the developing nature of the wage-labor system. The process of industrialization contrasted with that which unfolded in southern New England, whose coastal cities did not serve as points of debarkation for British immigrants. While textile capitalists rapidly adopted new machine technologies in spinning and weaving in rural New England, yarn and cloth production in the Philadelphia region remained exclusively in the hands of handloom weavers and mule spinners until relatively late. Only in the 1820s, with the expansion of an international market for coarse cloth, did the putting-out system, weaver's shop, and manufactory workhouse give way to the water-powered textile factory.

The fact that the new industrial capitalists, like J. J. Borie and Joseph Ripka, used a labor force of immigrant women and children goes a long way toward explaining the oppressive nature of production in the mills of Manayunk. Operating their factories in a competitive and crisis-ridden environment, Manayunk's mill owners created an arduous and unremitting system of labor around their machines, a system that was unregulated by the state or by moral compunction. These individual capitalists, in contrast to the early corporate mill owners of Lowell, oversaw a factory system of production that resembled that of the notorious Manchester mills.

Wage reductions, work conditions, and job displacement—what workers inside and outside of the mills measured as the inequities and injustices of the textile factory system—underlay the Luddism and mill strikes in the 1830s. Such conflict created what Philip Scranton has fittingly dubbed Manayunk's "tumultuous era," which had no parallel in the textile districts of southern New England. The "large scale confrontations between labor and capital" of the kind that overtook Manayunk in 1833 and 1834 were, as Jonathan Prude has concluded, "generally rare among antebellum textile operatives."[44]

The particular system of textile factory production in Manayunk was the basis for the confrontation between labor and capital in the early 1830s, but the oppositional ideology of the transatlantic textile worker provided the heightened solidarity that sustained the strikes and gave rise to a trade union movement among factory workers. Irish- and English-born textile workers were a mobile proletariat, who often came to the mills of Manayunk with years of experience in the mills of England and the Philadelphia region. They carried from the textile districts of Lancashire the radical ideas of British anti-Ricardians and trade unionists. New England textile workers judged early industrial capitalists and their mill system with the values and expectations of a "rural and artisan culture." Similarly, journeymen of the metropolitan and manufacturing centers shared a critique of capitalist relations grounded in indigenous "artisan republicanism."[45] Philadelphia and Manayunk mill workers, however, imported a body of beliefs concerning the rights of the modern wage laborer.

Historians have generally concluded that the American factory textile worker was essentially complacent during the rise of organized labor in the decade before the Panic of 1837. The transatlantic history of Philadelphia's cotton textile operatives and industry, however, suggests that we revise the interpretation that the early factory turnouts against wage reductions were defensive and thus inconsequential in the early labor movement. The factory system of textile production in Manayunk exhibited on a small scale the characteristics of England's cotton textile industry in the early decades of the nineteenth century—an industry that combined advanced technology with, as John Foster puts it, "the worst consequences of unplanned, competitive organization." Foster has noted that England's laboring class in this same period was cooperating and united as never before. Thus, "at one and the same time workers experienced a new strength and a new weakness." It was because of the historical convergence of these two developments, Foster argues, that England's "pioneer class consciousness" proved to be "peculiarly industrial in origin."[46]

The process of industrialization as it unfolded in Philadelphia and Manayunk did not merely parallel in miniature events in Britain; rather it was an extension of a social and economic transformation that began in England's northern textile districts before the end of the American Revolution. In and around America's major immigrant ports, Britain's mobile textile operatives were the agents of change in this transatlantic process. Given their role in Philadelphia, we must reevaluate our understanding of the origins of our "pioneer class consciousness" and consider that the "birth of nineteenth-century American working-class radicalism" occurred not solely within the skilled trade organizations of the indigenous urban artisan but also in the strike and union movements of Philadelphia's immigrant mill operatives.[47]

The new social relations of the factory system of production, the arguments posed by battling mill owners and operatives, and the often violent nature of their interaction underlay the development of Roxborough's early industrial society. In the late summer of 1833, "A Jeffersonian Workingman," the anonymous voice of the striking mill workers, spoke to the significance of the industrial strikes of the period for the whole community. "A spirit is at last awakened among the people of Manayunk," he wrote in response to the initial turnout by Joseph Ripka's workers in August. "A great change must inevitably take place, either for better or worse, as regards the manufacturing class of that village."[48] And indeed, over the subsequent four years, all groups in Roxborough struggled in the mills, the churches, and political parties over the fate of the "manufacturing class" and the "manufacturing aristocracy" in their community. The people of Roxborough participated in Jacksonian and Whig party politics and in the evangelical revivals and reform movement between 1832 and 1837 because they were concerned about the economic and political ruptures that accompanied the rise of the factories. They made wage reductions, trade unionism, evangelical conversion, temperance reform, "the Bank," and political alignments issues in the battle between labor and capital. Through their involvement in these struggles the people of the Manchester of America had played a vital role in the forming of industrial society.

NOTES

INTRODUCTION

1. *Democratic Press*, March 7, 1823; Samuel Hazard, ed., *Register of Pennsylvania* 2 (July–Jan., 1828): 14–15. Sean Wilentz discusses how mechanization and factory production were "virtually non-existent in New York" even by 1850. Sean Wilentz, *Chants Democratic: New York City and the Rise of the American Working Class, 1788–1850* (New York, 1984), pp. 112, 114.
2. Alan Dawley, *Class and Community: The Industrial Revolution in Lynn* (Cambridge, Mass., 1976); Thomas Dublin, *Women at Work: The Transformation of Work and Community in Lowell, Massachusetts, 1826–1860* (New York, 1979); Paul G. Faler, *Mechanics and Manufacturers in the Early Industrial Revolution: Lynn, Massachusetts, 1780–1860* (Albany, 1981); Susan E. Hirsch, *Roots of the American Working Class: The Industrialization of Crafts in Newark, 1800–1860* (Philadelphia, 1978); Bruce Laurie, *Working People of Philadelphia, 1800–1850* (Philadelphia, 1978); Jonathan Prude, *The Coming of Industrial Order: Town and Factory Life in Rural Massachusetts, 1810–1860* (Cambridge, Mass., 1983); Anthony F.C. Wallace, *Rockdale: The Growth of an American Village in the Early Industrial Revolution* (New York, 1980); Wilentz, *Chants Democratic*.
3. This is confirmed in Philip Scranton's fine study, *Proprietary Capitalism: The Textile Manufacture at Philadelphia, 1800–1885* (Cambridge, 1983).
4. Prude, *Industrial Order*, pp. xiii, 136, 142, 143, 225; Dublin, *Women at Work*, pp. 88, 107.
5. Wilentz, *Chants Democratic*, pp. 142, 213–14, 238; Faler, *Mechanics and Manufacturers*, chaps. 3 and 9, passim; Dawley, *Class and Community*, chap. 2, passim; Charles G. Steffen, *The Mechanics of Baltimore: Workers and Politics in the Age of Revolution, 1763–1812* (Urbana, Ill., 1984), pp. 276, 279. On Cincinnati see Steven J. Ross, *Workers on the Edge: Work, Leisure, and Politics in Industrializing Cincinnati, 1788–1890* (New York, 1985), chap. 3, passim. Dublin argues that even Lowell's female textile workers shared the world view of the American mechanic and thus protested to protect their rights as "daughters of freemen" (*Women at Work*, pp. 93–94).
6. This is Frederick Engels's phrase, quoted in E. P. Thompson, *The Making of the English Working Class* (New York, 1963), p. 191.
7. I regret that the sources have not allowed me to explore systematically the issue of gender in the development of radical thought and action among

Manayunk's mill workers. The particular experiences and sensibilities of female mill workers certainly impinged upon the nature of social conflict in industrializing society.
8. Engels quoted in Thompson, *The English Working Class*, p. 191. See also John Foster, *Class Struggle and the Industrial Revolution: Early Industrial Capitalism in Three English Towns* (London, 1974).
9. John Wade, *History of the Middling and Working Classes* (London, 1883); quoted in Asa Briggs, "The Language of 'Class' in Early Nineteenth-Century England," in Asa Briggs and John Saville, eds., *Essays in Labour History* (London, 1960), pp. 146–47.

CHAPTER 1 LABOR AND CAPITAL IN THE EARLY PERIOD OF MANUFACTURING

1. Gordon C. Bjork, "The Weaning of the American Economy: Independence, Market Changes, and Economic Development," *Journal of Economic History* 24 (Dec. 1964):545, 541–60 passim; Drew R. McCoy, *The Elusive Republic: Political Economy in Jeffersonian America* (Chapel Hill, N.C., 1980), p. 360. On the transition in republican attitudes on manufacturing, see McCoy, ibid., chap. 4; John F. Kasson, *Civilizing the Machine: Technology and Republican Values in America, 1776–1900* (New York, 1976), pp. 6–32; David J. Jeremy, "The British Textile Technology Transmission to the United States: The Philadelphia Region Experience, 1770–1870," *Business History Review* 44 (Spring 1973):32; and Robert D. Arbuckle, *Pennsylvania Speculator and Patriot: The Entrepreneurial John Nicholson, 1757–1800* (University Park, Pa., 1975), p. 139.

   The analysis that follows in this chapter is largely based on two manuscript collections: The John Nicholson Papers, General Correspondence, 1772–1819, Pennsylvania State Archives, Harrisburg, and the John Nicholson Letterbooks, 1791–1798, Historical Society of Pennsylvania, Philadelphia. The former, which contains Nicholson's incoming correspondence, is organized alphabetically by correspondent and then sub-arranged chronologically. I have not cited each letter that provided evidence for my argument. Rather, I refer the reader to the letters of the individuals mentioned in the chapter. Of particular importance is the correspondence between Nicholson and William Pollard, John Campbell, John Lithgow, Henry Elouis, Thomas Joubert, and William Eichbaum. Readers may also want to refer to my dissertation for identification of specific letters.
2. When Nicholson died at the age of forty-three he owed creditors 12 million dollars (Arbuckle, *Nicholson*, pp.2, 3, 139–42, and chap. 9). William Pollard to John Nicholson, April 27, 1793, John Nicholson Papers, General Correspondence, 1772–1819, MG 96, Pennsylvania State Archives, Harrisburg, Pennsylvania (hereafter cited as GC).
3. On Pollard's background and effort to patent the Arkwright frame, see Anthony Wallace and Davis J. Jeremy, "William Pollard and the Arkwright Patents," *William and Mary Quarterly*, 3d ser. (July 1977): 34:409–17; Samuel Batchelder, *Introduction and Early Progress of the Cotton Manufacture in the United States* (Boston, 1863), p. 35. See also Estimate of the Cost of Machinery . . . by William Pollard for John Nicholson, March 1, 1793, Samuel Wetherill Papers, Acc. 1436, Eleutherian

Mills–Hagley Foundation Library, Wilmington, Delaware (hereafter cited as Wetherill Papers).
4. Estimate of Cost of Machinery..., March 1, 1793, Wetherill Papers; Taylor to Nicholson, January 26, 1796, GC.
5. Pollard told Nicholson how farms proved advantageous to factory proprietors in England. For one thing "an advance price could be obtained within the town" (Pollard to Nicholson, May 17, 1795, GC). Nicholson did not keep up with George Stumme's demands for supplies, tools, and equipment, and Stumme returned to Germany frustrated by his lack of success.
6. Francis A. de La Rochefoucauld-Liancourt, *Travels through the United States of North America... 1795, 1796, and 1797* (London, 1799), 1:4.
7. Lithgow to Nicholson, Oct. 21, 1794, GC.
8. Memorial of Mr. Pallvison of Scotland on Labor Saving Machinery, n.d., Tench Coxe, Correspondence and General Papers, Historical Society of Pennsylvania (hereafter cited as Coxe Papers). Campbell also warned Nicholson that "a great deal of capital" was needed owing to the "difficulty of procuring machinery and workmen" (Campbell to Nicholson, April 4, 1795, GC).
9. Rochefoucauld-Liancourt, *Travels*, 1:5.
10. On wage-rate ratios of skilled to unskilled, see Donald R. Adams, "Some Evidence on English and American Wage Rates, 1790–1830," *Journal of Economic History* 30 (Sept. 1970):499–511, and Adams, "Wage Rates in the Early National Period, Philadelphia, 1785–1830," *Journal of Economic History* 28 (Sept. 1968):404–26.
11. Nicholson to Pollard, April 25, April 27, 1793, I, LB; Pollard to Nicholson, April 27, April 30, May 4, 1793, GC.
12. Lithgow to Nicholson, Oct. 21, 1794; April 8, 1795, GC.
13. Pollard to Nicholson, May 12, May 15, May 31, July 10, August 4, 1793; Oct. 10, 1794; May 17, 1795, GC. Pollard complained to Nicholson, "I have one man lately from London who understands every part of my Work, and I should be exceedingly sorry to turn him off least the Company at Paterson should get him." Pollard described his English and Scottish workers as "capital in their way" (Pollard to Nicholson, Sept. 24, 1793, GC).
14. Campbell was arrested in New York in his unsuccessful attempt. It is unclear what law he violated (Campbell to Nicholson, Nov. 3, 1794; April 10, 1795, GC). On another occasion, Nicholson had to bail Campbell out of jail for smuggling three textile workers out of Scotland (Arbuckle, *Nicholson*, p. 148).
15. Accounts of Workmen, Book A, 1794–1796, Book B, 1795–1797, Glass Works Accounts, John Nicholson Papers, Individual Business Accounts (hereafter cited as IBA); Thomas C. Cochran, ed., *The New American State Papers, Manufactures* (Wilmington, 1972), 1:73. On the skill of glass workers, see Joan Wallack Scott, *The Glassworkers of Carmaux: French Craftsmen and Political Action in a Nineteenth-Century City* (Cambridge, Mass., 1974).
16. Christian Triepel, Charles Eckhard, Francis Leitz, John Galder, and George Feunhackle to Nicholson, 19 April 1795; Christian Triepel and Charles Eckhard to John Nicholson, June 28, 1795, GC.
17. Henry Elouis to Nicholson, Feb. 13, 1796, GC.

18. For wage rates of boatmen and quarrymen, which varied between 71 and 78 cents (five and a half and six shillings) per day, see Account that Henry Elouis gave to John Nicholson..., IBA. On the wages of button workers, which varied from approximately $11.70 to $14.56 per month, see Ledger, 1795–1796, Button Works Accounts, IBA.
19. Delaware and Schuylkill Navigation Company, Rough Minutes, Nov. 10, 1792, Historical Society of Montgomery County, Norristown, Pennsylvania.
20. Workmen's Time Books, 1793–1794, Delaware and Schuylkill Navigation Canal Company Accounts, 1792–1795, IBA; ibid., Accounts and Receipts, 1792–1795; Delaware and Schuylkill Navigation Company, Rough Minutes, Dec. 3, 1794.
21. Pollard to Nicholson, Oct. 6, 1794, GC.
22. Nicholson to Eichbaum, June 15, 1795, II, LB; Jan. 8, 1796, III, LB; Nicholson to Thomas Bourne, Jan. 8, 1796, III, LB; Nicholson to William England, March 28, 1796, IV, LB.
23. Nicholson to Thomas Joubert, April 11, 1796, IV, LB; Elouis to Nicholson, March 5, 1795, GC.
24. Rochefoucauld-Liancourt, *Travels*, 1:5. The workmen were "not affording [him] much profit," Nicholson told Jacob Servoss (Nicholson to Servoss, April 14, 1795, I, LB).
25. On refusal to accept certain jobs, see Elouis to Nicholson, Feb. 2, 1796, GC; Eichbaum to Nicholson, May 11, 1796, GC.
26. Joubert to Nicholson, Aug. 11, 1796, GC. See also Nicholson to Joubert, March 31, 1796, Papers of John Nicholson, Box 2910, no. 1, Library of Congress; Elouis to Nicholson, Feb. 2, 1796, GC; Eichbaum to Nicholson, May 20, 1796, GC.
27. Eichbaum to Nicholson, Oct. 17, 1795, May 20, 1796, GC. On deducted time, see Accounts of Workmen, Book A, 1794–1796, Book B, 1795–1797, Glass Works, IBA. On glass manufacturing, see Arlene Palmer Schwind, "The Glass Makers of Early America," in *The Craftsman in Early America*, ed. Ian M. G. Quimby (New York, 1984), pp. 158–189. Joan Scott has described how molten glass was prepared from noon until midnight in the furnaces. The glass workers then blew bottles from midnight to noon when it was coolest (*Glassworkers of Carmaux*, p. 32).
28. Eichbaum to Nicholson, Jan. 29, 1796, GC. On one occasion, Elouis relayed the request of the carpenters who were "raising a new home" for "ten gallons of spirits" (Elouis to Nicholson, March 22, 1795, GC). Joubert requested liquor—"a few gallons of anything"—to make sure the glass workers finished an order. (Joubert to Nicholson, May 20, 1796, GC). On drinking as a custom and practice of premodern work life, see Howard B. Rock, *Artisans of the New Republic: The Tradesmen of New York City in the Age of Jefferson* (New York, 1979), pp. 296–97; W. J. Rorabaugh, *The Alcoholic Republic: An American Tradition* (New York, 1979), pp. 132–33; and Scott, *Glassworkers of Camaux*, p. 42.
29. Elouis to Nicholson, Jan. 20, 1795, GC; N. F. Mix to Nicholson, Jan. 5, 1795, GC.
30. Elouis to Nicholson, Feb. 2, 1796, GC.
31. Herbert Gutman, *Work, Culture and Society in Industrializing America* (New York, 1977), p. 19. See also E. P. Thompson, "Time, Work-Discipline, and Industrial Capitalism," *Past and Present* 38 (Dec. 1967): 56–97.

32. Karl Marx, *Capital, A Critique of Political Economy* (1867; rpt. New York, 1977), 1:456–57; Maurice Dobb, *Studies in the Development of Capitalism* (London, 1946), p. 259; Sidney Pollard, *The Genesis of Modern Management: A Study of the Industrial Revolution in Great Britain* (London, 1965), pp. 65–66; Laurie, *Working People*, p. 13.
33. My definition of the essential nature of the manufactory vis-à-vis handicraft production and factory production differs from that of Sean Wilentz (*Chants Democratic*; p. 115).
34. Jonathan Prude describes similar dynamics in the tension existing in Samuel Slater's Pawtucket mill in the 1790s (*Industrial Order*, pp. 44–45). Through most of the late eighteenth century, the income of Philadelphia laboring families barely matched the cost of basic necessities. During the early 1790s, real wages fluctuated 20 to 25 percent below 1762 levels and posed severe problems for those at the lower end of the economic order. See Billy G. Smith, "The Material Lives of Laboring Philadelphians, 1750–1800," *William and Mary Quarterly*, 3d ser. (April 1981) 38:188. Donald Adams, Jr., has shown that real wages declined for artisans and laborers between 1794 and 1796, precisely the period that Nicholson employed manufactory workers. While the price of labor was high, so was the cost of living. See Adams, "Wage Rates," table 1, p. 406; table 9, p. 425, and Smith, "Laboring Philadelphians," pp. 188, 201–2.
35. Rochefoucauld-Liancourt, *Travels*, 1:5; Nicholson to Servoss, April 14, 1795, I, LB.
36. Nicholson had switched to a system of piece rates, paying the quarrymen per perch of stone in the summer. On discontent of other laborers, see also Elouis to Nicholson, Oct. 15, Dec. 5, 1795, GC. The boatmen had all quit by January 1796. On the boatmen's discontent, see James Lovett to Nicholson, Dec. 17, 1795, GC; Elouis to Nicholson, April 11, 1795; March 18, 1796, GC.
37. Pollard to Nicholson July 14, 1795. A few workers left to search for work in Wilmington and New York. Pollard to Nicholson, July 20, July 25, Aug. 25, Aug. 28, Sept. 1, Sept. 16, 1795, GC; Nicholson to Pollard, Sept. 28, Sept. 30, 1795, III, LB.
38. Pollard's sensibilities and his particular relationship with his workers parallels that of his contemporary Samuel Slater in the Almy, Brown, and Slater enterprises in Pawtucket (Prude, *Industrial Order*, p. 45).
39. Mix blasted Nicholson: "There is no reason but one that you have not cleared fifteen hundred pounds[;] that is you have not done your part" (Jonathan Mix to Nicholson, Feb. 28, 1796, GC; see also Statement of Quantity of Buttons that Might be made in a Year, Button Works Accounts, IBA; Thomas Bourne to Nicholson, March 16, 1796, GC).
40. The button workers' wages were approximately $11.70 to $14.56 per month, Ledger, 1795–1796, Button Works Accounts, IBA.
41. Charles V. Hagner, *Early History of the Falls of Schuylkill, Manayunk, Schuylkill, and Lehigh Navigation Companies, Fairmont Waterworks, Etc.* (Philadelphia, 1869), pp. 33–34.
42. See John Nicholson, Esq., Day Book for the Button Manufactory Began November 13, 1794, Button Works Accounts, IBA; ibid., Ledger, 1795–1796.
43. Jonathan Mix to Nicholson, Feb. 28, 1796, GC. Mix's nephew, Nathaniel, who was employed by Nicholson, informed on his uncle's practice of selling boxes of buttons (N. F. Mix to Nicholson, May 2, 1796, GC). See

also Button Works Accounts, Ledger, 1795–1796; Nicholson to Dorris Higgens, Samuel Dennison, and Ashbile Baker at the Button Works, 23 May 1797, VII, LB.
44. Flood to Nicholson, June 10, 1795; April 9, 1796, GC.
45. See Taylor and Bowler to Nicholson, July 7, 1795, GC; Campbell to Nicholson, March 7, April 10, 1795, GC; John McGann, James Robertson, John Hailet, D. Lamb, John Reed, James Robertson, Jr., to Campbell, April 6, 1795, GC; Nicholson to Servoss, April 14, 1795, I, LB.
46. Joan Scott argues that the skill of the master glassblower in nineteenth-century France (the *souffleur*) put him in a powerful position in relationship to his employer (*Glassblowers of Carmaux*, p. 68). On the steps of production in bottle-making performed by a team of glass workers, see ibid., pp. 23–31.
47. Eichbaum to Nicholson, July 9, 1795, GC. On output, see Accounts of Workmen, Book A, 1794–1796, Book B, 1795–1797, Glass Works Accounts, IBA; Records of Glass Made, March, May, 1797, Glass Works Accounts, IBA. Dismayed on one occasion when the pots broke and he could not meet an order for 15,000 claret bottles, Nicholson wrote Eichbaum, "It is a mortifying circumstance that the work should be stopped when the demand is so great" (Nicholson to Eichbaum, Feb. 2, 1796, III, LB).
48. Mathew Fertner to Nicholson, July 19, 1795, GC. Fertner died two days after writing the letter.
49. Nicholson to M. Kepple, Dec. 14, 1795, III, LB.
50. Joubert to Nicholson, June 15, 1796, GC.
51. Gutman, *Work, Culture, and Society*, p. 19.
52. Nicholson to Eichbaum, March 4, 1797, VI, LB.
53. Joubert to Nicholson, June 2, 1797, GC; Nicholson to Joubert, June 2, June 3, 1797, VII, LB. Nicholson had tried desperately to keep the glass house producing despite the fact that the sheriff had put a levy on the property and tools. At the end of May, the sheriff began to seize the movable materials of Nicholson's glass manufactory, including the personal property of the supervisors (Joubert to Nicholson, May 25, 1797, GC). Joubert was running the complex for Nicholson and had attempted to pare down costs in the other branches, closing the stocking and button manufactories, foundry, and store. Nicholson tried to keep the boatmen transporting the glass bottles to prevent the sheriff from seizing them (Nicholson to Joubert, May 26, May 29, May 30, 1797, VII, LB).
54. Flood to Nicholson, June 15, 1795, GC.
55. Nicholson complained to Pollard that Bowler "hath purloined my property and after all the money paid him has taken passage, and off with the plunder" (Nicholson to Pollard, Sept. 11, 1795, III, LB).
56. He informed Pollard that he was also "just about at sixes and sevens" with him (Nicholson to Pollard, Sept. 7, 1795, III, LB).
57. Jane Eichbaum to Nicholson, June 9, 1797, GC.
58. Taylor to Nicholson, April 6, 1797, GC. Taylor was indeed penniless. In March he wrote Nicholson to send him money that night for "I am summoned for five [dollars] which I must pay tomorrow morning before seven o-clock in the morning or go to jail" (Taylor to Nicholson, March 22, 1797, GC).
59. Taylor to Nicholson, April 11, 1798, GC.
60. England to Nicholson, Sept. 5, 1797, GC.

61. Lithgow to Nicholson, May 11, 1799, GC; Wallace and Jeremy, "William Pollard," p. 424.
62. See Arbuckle, *Nicholson*, chap. 9.

CHAPTER II Textiles and the Urban Laborer, 1787–1820

1. The New England experience has been meticulously reconstructed by Caroline F. Ware, *The Early New England Cotton Manufacture: A Study in Industrial Beginnings* (New York, 1931); Dublin, *Women at Work*; Prude, *Industrial Order*; and Gary Kulik, "Pawtucket Village and the Strike of 1824: The Origins of Class Conflict in Rhode Island," *Radical History Review* 17 (Spring 1978):5–37.
2. New York, while a busier immigrant post than Philadelphia, was neither a center of handloom weaving or of cotton yarn production (Wilentz, *Chants Democratic*, pp. 31, 169).
3. Gary B. Nash, *The Urban Crucible: Social Change, Political Consciousness, and the Origins of the American Revolution* (Cambridge, Mass., 1979), pp. 189–92; and see Nash, "The Failure of Female Factory Labor in Colonial Boston," *Labor History* 20 (Spring 1979):165–88. See also Victor S. Clark, *History of Manufactures in the U.S.* (Washington, D.C., 1921), 1:188, and McCoy, *The Elusive Republic*, pp. 116–17.
4. The manufactories of the 1760s, like the earlier Boston venture, could not compete in quality or price with British imports. Nash, *Urban Crucible*, pp. 225, 327, 332, 335.
5. See letters in Dunlap's *Pennsylvania Packet and General Advertizer*, Oct. 16, Dec. 4, 1775, quoted in William R. Bagnall, *The Textile Industries in the United States*, (Cambridge, 1893; rpt. New York, 1971), 1:69–70.
6. Tench Coxe, *An Address to an Assembly of the Friends of American Manufactures*, 1787, pamphlet, Eleutherian Mills–Hagley Foundation Library, p. 23. The society did not obtain any public money as support until March of 1789, when the Pennsylvania legislature approved the purchase of $1,000 worth of shares (Bagnall, *Textile Industries*, 1:78; and see Arbuckle, *Nicholson*, pp. 139–40).
7. Henry Simpson, *The Lives of Eminent Philadelphians Now Deceased* (Philadelphia, 1859), pp. 942–43.
8. On the advantages of the putting-out system, see Pollard, *Modern Management*; p. 32, and David S. Landes, *The Unbound Prometheus: Technological Change and Industrial Development in Western Europe from 1750 to the Present* (Cambridge, 1969), pp. 56, 118–19.
9. Pennsylvania Society for the Encouragement of Manufactures and the Useful Arts, Minutes of the Manufacturing Committee, Historical Society of Pennsylvania, Philadelphia, vol. 1, Dec. 21, 1787; Jan. 22, 1788 (hereafter cited as PSEMUA). The most that any woman made in one month was $5.52, which by itself was insufficient even as a subsistence (Manufacturing Society of Philadelphia, 1788–1789, Account Book, Library of Congress, Washington, D.C.). The sources do not provide evidence about the household situations or livelihoods of individual outworkers.
10. The contemporary account of the Federal Procession is in the *American Museum* and is reprinted in Bagnall, *Textile Industries*, 1:110.
11. From a letter of 1809 quoted by Ware, *New England Cotton Manufacture*, p. 51. On centralizing production and embezzlement, see Harry Braver-

man, *Labor and Monopoly Capital: The Degradation of Work in the Twentieth Century* (New York and London, 1974), p. 63, and Pollard, *Modern Management*, p. 33.
12. PSEMUA, vol. 2, May 6, 1789.
13. PSEMUA, vol. 1, April 29, 1789. The factory weavers collectively presented wage demands to the board of manufacturers (Ibid., Oct. 8, 1788, May 5, 1789; Bagnall, *Textile Industries*, 1:110).
14. Bagnall, *Textile Industries*, 1:78; PSEMUA, vol. 1, Sept. 27, Nov. 24, 1787; Jan. 22, Dec. 6, April 2, 1788.
15. The spinning jenny required that the operator turn a wheel that drove up to eighty spindles and gave twist to the yarn. A clasp attached to a carriage replicated the spinner's hand in stretching the thread to a proper length. When the yarn reached the right twist and size, the spinner would return the carriage to the frame, performing the spinner's "second process" of winding the yarn on a bobbin ready for a weaver's shuttle. See account from *Edinburgh Encyclopedia*, quoted in George S. White, *Memoir of Samuel Slater: The Father of American Manufacture* (Philadelphia, 1836), p. 69; James Montgomery, *The Theory and Practice of Cotton Spinning; or The Carding and Spinning Master's Assistant* (Glasgow, 1833), p. 151; Andrew Ure, *The Cotton Manufacturer of Great Britain Investigated and Illustrated* (London, 1861), 1:226–29; and Ware, *New England Cotton Manufacture*, p. 24.
16. J. L. Hammond and Barbara Hammond, *The Skilled Labourer, 1760–1832* (London, 1919), pp. 50, 53; S. J. Chapman, *The Lancashire Cotton Industry* (Manchester, 1904), p. 54; Montgomery, *Cotton Spinning*, p. 69.
17. Bagnall, *Textile Industries*, 1:79.
18. PSEMUA, vol. 1, Sept. 29, Oct. 3, 1787; July 9, April 23, April 30, Sept. 24, 1788; vol. 2, Feb. 18, April 2, 1789. The Guardians of the Poor had put their inmates to work on indoor spinning wheels and needed the room for their own production. See Guardians of the Poor, Daily Occurrence Dockets, City and County Archives of Philadelphia, Philadelphia, Jan. 6, April 5, 1790. Early textile manufacturers in Pawtucket, in contrast, faced a labor scarcity (Prude, *Industrial Order*, p. 43).
19. In the spring of 1789, the General Assembly of Pennsylvania passed a law to assist cotton manufacturers of the state (PSEMUA, vol. 1, April 2, May 6, 1789; vol. 2, May 21, 1789).
20. *Freeman's Journal*, April 14, 1790.
21. Chapman, *Lancashire Cotton Industry*, p. 76; Hammond and Hammond, *The Skilled Labourer*, p. 54. The Hammonds write that the jenny caused riots when introduced in 1776 in the southwest of England (p. 145). Andrew Ure wrote that because of "an enraged populace, who even threatened his life, [Hargreaves] migrated to Nottingham." (Ure, *Cotton Manufacturer*, p. 228; Montgomery, *Cotton Spinning*, p. 152).
22. Tench Coxe, February 1804, Eleutherian Mills–Hagley Foundation Library.
23. Memorial of Mr. Pallvison of Scotland on Labor Saving Machinery, Coxe Papers. Maurice Dobb discusses how eighteenth-century inventors of labor-saving machines were conscious of the desire of manufacturers to save the cost of labor and thus enhance their profit. See *Studies in the Development of Capitalism* (London, 1946), p. 276.

24. See advertisement of James Davenport, Fall 1797, quoted in Bagnall, *Textile Industries*, 1:225.
25. Alexander Hamilton, "Prospectus of the Society for Establishing Useful Manufactures" (1791), in Arthur Cole, ed., *Industrial and Commercial Correspondence of Alexander Hamilton* (Chicago, 1928), p. 192.
26. Coxe, *An Address to . . . American Manufacturers*, pp. 89.
27. "Thoughts on . . . the manufactory of superfine woolen cloths," Juriscola, n.d., Coxe Papers. One supporter of Hamilton's plan to establish a cotton spinning and cloth factory in New Jersey found out for him that the wages of women with boarding and lodging were less than half that of ablebodied men in the area. (Thomas Lowrey to Hamilton, Oct. 14, 1971, in Cole, *Hamilton*, p. 220).
28. Coxe, *An Address to . . . American Manufacturers*, pp. 89; Alexander Hamilton, "Report on Manufacturers, December 1791," in Thomas C. Cochran, ed., *The New American State Papers, Manufactures*, (Wilmington, 1972), 1:47. Hamilton received detailed outlines for launching the Society for Useful Manufactures' own spinning mills, one of which noted that the mill "almost entirely avoids manual labor" ("Estimate of the Cost of Equipping a Flax and Hemp Spinning Mill," in Cole, *Hamilton*, pp. 225–28).
29. Hamilton, "Report on Manufactures, December 1791." On the concern to make the poor useful through manufactures, see McCoy, *The Elusive Republic*, p. 116.
30. Edith Abbott, *Women in Industry: A Study in American Economic History* (New York, 1900), p. 43; Juriscola, "Thoughts on . . . the manufactory of superfine woolen cloths," Coxe Papers.
31. Maldwyn A. Jones, "Ulster Emigration, 1783–1815," in E. R. Green, ed., *Essays in Scotch-Irish History* (New York, 1969), pp. 49, 52, 63; Clark, *Manufacturers in the U.S.*, p. 400. Billy G. Smith, "Struggles of the 'Lower Sort': The Lives of Philadelphia's Laboring People, 1750–1800" (Ph.D. diss., UCLA, 1981), pp. 57, 72, and n. 79. Edward Carter has calculated that Irish immigration averaged about three-thousand persons annually for the balance of the 1790s. See "A 'Wild Irishman' under Every Federalist's Bed: Naturalization in Philadelphia, 1789–1806," *Pennsylvania Magazine of History and Biography* 94 (July 1970):33.
32. Jones, "Ulster Emigration," p. 52; William Forbes Adams, *Ireland and Irish Emigration to the New World from 1815 to the Famine* (New Haven, 1932), pp. 50–51; L. M. Cullen, "The Irish Economy in the Eighteenth Century," in L. M. Cullen, ed., *The Formation of the Irish Economy* (Cork, 1968), p. 30.
33. Chapman, *Lancashire Cotton Industry*, p. 53; Hammond and Hammond, *The Skilled Labourer*, p. 52. The Arkwright frame, completely powered by water, entirely eliminated the skilled spinner from the labor process. The only skill needed for tending this machine was the ability to knot a broken thread (McMurray, "Technological Change," pp. 40–42; Montgomery, *Cotton Spinning*, p. 155; Ware, *New England Cotton Manufacture*, p. 26).
34. By 1800 weavers had organized societies to protest the degraded price of their labor (Hammond and Hammond, *The Skilled Labourer*, pp. 42; 57–58, 153; Thompson, *The English Working Class*, p. 278).
35. To the Manager or Partners of a Company for Weaving Cotton Cloth late

got up or Erected in Philadelphia, July 6, 1790, in Cole, *Hamilton*, pp. 183–84; see also Chapman, *Lancashire Cotton Industry*, p. 38; and William Pollard to John Nicholson, April 30, May 4, April 27, 1793, GC.

36. Charles Janson, *A Stranger in America, Containing Observations Made during a Long Residence in the Country, on the Genius, Manners, and Customs of the People of the United States* (London, 1807), p. 452.
37. Janson, *A Stranger in America*, p. 452.
38. Byrnes arrived "almost nak'd" (Guardians of the Poor, Daily Occurrence Dockets, Sept. 8, Oct. 18, 1792; *Gale's Independent Gazette*, Jan. 3, 1797, quoted in John Alexander, *Render them Submissive: Responses to Poverty in Philadelphia, 1760–1800* [Amherst, 1980], p. 78). The Passenger Act set up a quota for the number of passengers by the tonnage of the ship. This so limited the amount of passengers that merchants more than tripled fares to make up for their losses (Jones, "Ulster Emigration," p. 58). Edward Carter has figures that one-third of Philadelphia's Irish immigrants in the 1790s remained in the seaport ("Naturalization in Philadelphia," p. 343).
39. Quotations are from the constitution of the Hibernian Society printed in the *Freeman's Journal*, March 31, 1790. See also Carter, "Naturalization in Philadelphia," p. 332; Erna Risch, "Immigrant Aid Societies Before 1820," *PMHB* 60 (Jan. 1936):30
40. Ware, *New England Cotton Manufacture*, pp. 27, 23; Dublin, *Women at Work*, p. 16.
41. Ware, *New England Cotton Manufacture*, p. 28; Clark, *Manufactures in U.S.*, p. 387; Rola Tryon, *Household Manufactures in the United States, 1640–1860: A Study in Industrial History* (Chicago, 1917), p. 37.
42. James Davenport for a short time manufactured sailcloth using a water frame to spin the flax (Bishop, *American Manufactures*, 1:72). Samuel Wetherill, the veteran textile manufacturer, drew up a grand blueprint to duplicate "the cotton manufacture at Manchester" in his city. He foresaw the mill housing 16,000 spindles (Samuel Wetherill, "To the Publick," Coxe Papers).
43. Ware, *New England Cotton Manufacture*, p. 209.
44. On the technology of the mule and the skill necessary to operate it, see Montgomery, *Cotton Spinning*, pp. 167–70.
45. U.S. Congress, House, 2nd Sess. Robert Gallatin, "Report on Manufactures," April 1810, in Cochran, *New American State Papers, Manufactures*, 1:125–26; Tench Coxe, *A Statement of the Arts and Manufactures of the United States of America for the Year 1810* (Philadelphia, 1814), p. 24. Gallatin's report estimated that 3,550 women and children were employed in the factories that contained the labor-saving machinery (pp. 125–26).
46. Coxe, *Statement of . . . Manufactures*, 1814, p. 24.
47. Janson, *A Stranger in America*, pp. 195–96.
48. Quoted in Arthur Redford, *Labor Migration in England, 1800–1850* (Manchester, England, 1926), p. 21.
49. Pollard, *Modern Management*, pp. 162–63, 310 n. 2; and see McCoy, *The Elusive Republic*, p. 108.
50. The discussion on the Guardians of the Poor that follows is based largely on the Minutes of the Manufacturing Committee (hereafter cited as GP, Minutes) and the Weekley Statement of Persons Employed in the Man-

ufactory... 1807 (hereafter cited as GP, Weekly Statement). Both are part of the Committee on Manufacturing, Minutes, 1807–1887, Guardians of the Poor (Record Group 35), Archives of the City and County of Philadelphia.

51. The physical layout of the factory was simple, each step of cloth production being located in a separate room or floor. Within the factory, for example, cotton cloth from the West Indies or the Sea Islands was cleaned and picked on the ground floor, carded and spun on wheels, mules, or jennies in different departments, woven on handlooms in the cellar, and bleached on the green next to the almshouse (GP, Minutes Sept. 2, March 25, 1807; April 26, Feb. 17, 1808).

52. GP, Minutes, Feb. 25. On stocking weavers, see GP, Minutes, September 1808. In January 1808 the accounts listed eight stocking knitters employed in the almshouse manufactory (GP, Minutes, Weekly Statements, January 1808). Bagnall described the Germantown hosiers as in a miserable state in 1804, (*Textile Industries*, pp. 362–63).

53. The distinctions between paupers and hired labor and between indoor and outdoor workers were clear. Paupers were inmates of the almshouse—residents of the county who could not independently maintain themselves. They were not paid a regular wage in exchange for their labor. These spinners and weavers, as well as the wage-earning weavers who boarded at the almshouse, the mule spinners, and other "tradesmen" who worked within the factory, were indoor workers. Outdoor workers referred to female wheel spinners and male handloom weavers who produced yarn and cloth in their homes within the putting-out system operated by the Guardians of the Poor. The terms "indoor" and "outdoor" were used consistently by the manufacturing committee and supervisors to distinguish the two groups of female wheel spinners.

In general, the textile labor force consisted of fewer paupers than hired factory and domestic outworkers. Before 1812 the proportion of indoor to outdoor female spinners fluctuated from month to month, although outdoor spinners usually outnumbered indoor spinners. The War of 1812 probably left many lower-class women without husbands and unable to subsist alone. After March of 1812 the number of outdoor spinners dropped below thirty and until 1815 (when the enumeration of indoor spinners stopped), the indoor spinners numbered in the fifties and sixties during the winter. Hired weavers consistently outnumbered pauper weavers.

54. Mule spinners could earn much more than weavers. Charles McCurdy, for example, received a monthly wage of twenty-five dollars plus three dollars per week for board. He was clearly a skilled worker who pleased the committee and was placed in charge of the spinning department in 1808 (GP, Minutes, Aug. 19, Sept. 30, Oct. 14, Oct. 28, 1807).

55. The number of laborers on the accounts increased dramatically; the work force of carding and roving attendants even tripled (GP, Minutes, Oct. 21, Oct. 10, 1807; June 7, 1809; Sept. 14, 1808; Dec. 4, 1811; Feb. 7, 1808).

56. Why the numbers plunged abruptly is unclear. It may have been a response to the repeal of the Embargo Act in March 1809 and its replacement with the less restrictive Nonintercourse Act.

57. The matron had charge of spinners, winders, and spoolers—all females. A mule spinner, Charles McCurdy, was placed in the position of supervisor of the carding and spinning and was in charge of delivering the yarn

to the weavers' supervisor. The committee also employed an overseer of the yarn room to sort the yarn for the weavers and prevent any unsupervised persons from coming into the room (GP, Minutes, Feb. 10, Dec. 28, 1808).

58. Ibid., Sept. 25, 1811; June 17, Sept. 30, 1807; May 6, 1810.
59. Promoters of manufacturing advocated spending the "trifling but just and necessary expense of teaching [children] to read and write" (*Democratic Press*, July 2, 1807).
60. GP, Minutes, March 23, 1808; May 16, 1810.
61. The committee often found this to be a poor decision. They directed, for example, James Clark Graham, a resident weaver, to work on the spinning machinery—a billy and jenny. After the end of two months they concluded that his weekly output was too low and ordered him back to the loom (GP Minutes, Weekly Statement, Feb. 28, April 25, 1810).
62. The average weaver's output was about thirty yards per week of tow linen, sheeting or linsey at seven to thirteen cents per yard (ibid., Sept. 1811–Sept. 1812; Clark, *Manufactures in the U.S.*, p. 388).
63. GP, Minutes, Feb. 25, 1807. The committee stipulated that the superintendent would inspect all the cloth manufactured and "make such deductions for bad work, as to him may appear just" (ibid., Feb. 28, 1809).
64. Mule spinners also bargained with the committee in a shifting market. William Fletcher and John Batley, for example, were hired in June 1807 when there was a "deficiency in cotton yarn" and fired four months later "in consequence of bad conduct" when the committee found a spinner willing to work for less wages (GP, Minutes, June 17, Sept. 2, Aug. 19, Sept. 30, Oct. 14, Oct. 28, 1807).
65. For John Store, see ibid., Jan. 17, 1810; July 17, July 3, Nov. 13, 1811. For Andrew Nelson, see ibid., Dec. 28, 1808; Weekly Statements, Sept. 1811–Sept. 1812 and 1813. For Conrad Snyder, see ibid., Weekly Accounts, Sept. 1811–Sept. 1812; GP, Daily Occurrence Dockets, Nov. 17, 1812. A few weavers did make a steady and decent competency as employees of the almshouse factory. David Armstrong, for example, labored eight years on the factory looms as a hired weaver. Samuel Walker, who also appeared in the almshouse census in 1807, moved into the position of foreman of the factory in 1816 after nine years as a hired weaver (GP, Almshouse Census, 1807; GP, Minutes, Weekly Statements, Sept. 1811–Sept. 1812, 1815).
66. Redford, *Labour Migration*, pp. 19, 21–22.
67. GP, Minutes, Jan. 29, 1812.
68. GP, Minutes, Weekly Statement, Feb., March, 1811, especially Feb. 27, March 6.
69. GP, Minutes, Sept. 30, 1807; Nov. 6, 1811. In another case the committee accused three pauper laborers of stealing linen from the bleaching field (ibid., Dec. 2, 1807). On embezzlement, see Braveman, *Labor and Monopoly Capital*, p. 63, and Pollard, *Modern Management*, p. 33.
70. The committee ordered the paupers confined "in the cells" for a weekend and the hired weavers fired (GP, Minutes; March 2, 1808). The supervisors were ordered to discharge those who did not meet the daily schedule (ibid., June 17, 1811).
71. Louis Sears, "Philadelphia and the Embargo of 1808," *Quarterly Journal of Economics* 35 (Feb. 1921):356; Ware, *New England Cotton Manufac-*

ture, p. 48; Batchelder, *Cotton Manufacture*, p. 64.
72. Almy and Brown to E. W. Lawton, March 11, 1814, quoted in Ware, *New England Cotton Manufacture*, p. 48. For census figures, see table reprinted in Samuel Hazard, ed., *Register of Pennsylvania* 4 (Sept. 1828):165.
73. Ware, *New England Cotton Manufacture*, p. 54; Bishop, *American Manufactures*, p. 188.
74. Four girls or women could operate the machines and produce with the smallest six-spindle machine six pounds of cotton yarn in a fourteen-hour day (Letter by John G. Baxter, *Niles' Weekly Register*, March 5, 1814; Bishop, *American Manufactures*, p. 88).
75. Extract from letter from New Jersey to a friend in Philadelphia, *Niles' Weekly Register*, Supplement, March 1813–1814.
76. Chapman, *Lancashire Cotton Industry*, p. 44; Hammond and Hammond, *The Skilled Labourer*, p. 88; Jones "Ulster Emigration," p. 62.
77. *Niles' Weekly Register*, Aug. 17, 1816. And see Scranton, *Proprietary Capitalism*; p. 94.
78. Adams, *Ireland and Irish Emigration*, p. 98. Niles recapitulated the statistics from the passenger vessels' registers. See *Niles' Weekly Register*, July 27, Aug. 17, Oct. 5, 1816; May 31, June 21, Aug. 2, 1817; Jan. 10, Aug. 1, 1818.
79. Ibid., May 30, July 27, 1816. On immigrants remaining in the city, see Adams, *Ireland and Irish Emigration*, pp. 97–98, 106–7.
80. Adams, *Ireland and Irish Emigration*, p. 107
81. According to Edith Abbott, *Niles' Weekly Register* was filled with the postwar manufacturers' assurances that men would be left on the plough while women and children attended the new machines (*Women in Industry*, pp. 51–52). See also U.S., Congress, House, Committee of Commerce and Manufacturing, "Report of Committee . . . to which was referred a memorial and petitions to the manufactures of cotton wool . . . 1816," Eleutherian Mills–Hagley Foundation Library.
82. Adams, *Ireland and Irish Emigration*, table 10, p. 423; Thompson, *The English Working Class*, p. 279. To operate a handloom required the knowledge and skill to perform a number of different tasks. While the steps of the labor process could be quickly learned, dexterity and judgment separated masters from apprentices. By 1813 two heddles and one shuttle on the earliest powerlooms had virtually eliminated the need for a skilled weaver. On the new powerloom, the three basic motions of weaving were mechanized and imparted by waterpower (Arthur Harrison, in Cole, *The American Wool Manufacture* [Cambridge, Mass., 1926], p. 125; Gibb, *Saco-Lowell Shop*, p. 739). For a different interpretation of the beginnings of industrialization in the Philadelphia region, see Carville Earle and Ronald Hoffman, "The Foundation of the Modern Economy: Agriculture and the Costs of Labor in the United States and England, 1800–1860," *American Historical Review* 85 (Dec. 1980): 1055–94.
83. The number increased from 1,761 to 2,325 (Hazard, *Register of Pennsylvania* 4 [Sept. 1828]:168). For an interesting interpretation of the slimly documented outwork handloom weaving system ca. 1820, see Scranton, *Proprietary Capitalism*, pp. 83–86.
84. United States, Schedule of Manufactures, Fourth Census of the United States, 1820, Philadelphia County, Reel 14, National Archives, Washington, D.C. (hereafter cited as SM, 1820). David J. Jeremy has figured

that Pennsylvania represented only 6.7 percent of the national fixed capital investment in industrial cotton manufacture (*Transatlantic Industrial Revolution: The Diffusion of Textile Technology between Britain and America* [Cambridge, Mass., 1981], pp. 164, 276). Scranton also analyzed the 1820 manufacturing census. For a different and valuable treatment of this source, see *Proprietary Capitalism*, pp. 95–121.

85. See, for example, the schedules of Edward Erving, Alex Moffet, Thomas Grim, James Maxwell, and James Stanaghan, SM, 1820; Jeremy calls Philadelphia the exclusive location of "small, horizontally specialized cotton weaving firms" in 1820 (*Transatlantic Industrial Revolution*, p. 165).

86. The amount of capital invested in these spinning and weaving factories ranged from $8,000 to $130,000. No shop devoted exclusively to weaving had more than $5,000 invested in looms and buildings.

87. See Ware, *New England Cotton Manufacture*, p. 71, and Dublin, *Women at Work*, p. 18. A Committee of Manufacture of Cotton and Wool in the City of Philadelphia sent a memorial to Congress outlining their plight in January 1816. See Cochran, *New American State Papers, Manufactures*, p. 439.

88. The number of hands dropped from 2325 to 149 (Hazard, *Register of Pennsylvania* 4 [Sept. 1828]:168). See also Scranton, *Proprietary Capitalism* tables 4, 8, p. 21.

CHAPTER III MECHANIZATION AND MILL PRODUCTION IN THE URBAN SEAPORT, 1820–1837

1. Edwin T. Freedly, *Philadelphia and Its Manufactures: A Handbook Exhibiting the Development, Variety, and Statistics of the Manufacturing Industry of Philadelphia in 1857* (Philadelphia, 1858), pp. 233–34.
2. Report of the Managers of the Schuylkill Navigation Co. to the Stockholders, Jan. 7, 1822, Pennsylvania Transportation, Schuylkill Navigation Co., 3:7, no. 10, Historical Society of Pennsylvania; Reuben Haines to Samuel Batchelder, Hamilton Mills, Lowell, Oct. 16, 1830, Reuben Haines Correspondence, Wyck House, Germantown.
3. For a list of purchasers of water power, see Hagner, *Early History*, pp. 76–78; Thomas Wilson, *Picture of Philadelphia for 1824* (Philadelphia, 1823), pp. 10–11; Horatio Gates Jones in the *Manayunk Star*, April 10, 1855.
4. *Mechanics' Free Press*, Aug. 2, 1828; Nicholas B. Wainwright, "The Diary of Samuel Breck, 1823–1827," *Pennsylvania Magazine of History and Biography* 103 (Jan. 1979):102; Wainwright, "Diary of Breck, 1827–1833," *PMHB* 103 (April 1979):228; F. Murphy to Horatio J. McClean, Feb. 7, 1828, Bost. MS A398, Boston Public Library, Boston, Mass.
5. Hazard, *Register of Pennsylvania* 2 (July–Jan. 1828):14–15; *Niles' Weekly Register*, Dec. 1, 1827. Contrast the number of hands employed in five mills in 1828 to Jonathan Prude's findings for his communities where in 1832, 816 were employed in about a dozen mills (*Industrial Order*, p. 84).
6. *Germantown Telegraph*, Jan. 1, 1834; Pennsylvania, General Assembly, *Journal of the Senate, Session of 1837–38* (Harrisburg, 1838), 1:325, 2:326, 280–81 (hereafter cited as *PJS*). The Manayunk mills produced $748,900 worth of manufactured articles, 23 percent of the county total (U.S., Schedule of Mines, Agriculture, Commerce, Manufactures, etc., Sixth Census of the United States, 1840, Pennsylvania, Roxborough). William

A. Sullivan gives the figures for the County of Philadelphia in *Industrial Worker*, p. 19. For an overview of Philadelphia textile manufacturing and proprietary capitalism at mid-century and an explicit comparison to Lowell, see Philip Scranton, *Proprietary Capitalism*, pp. 178–95.
7. Clark, *History of Manufactures*, pp. 544–45; Ware, *New England Cotton Manufacture*, pp. 169, 310. "Cotton yarn as fine as #20 and goods out of such yarn is protected," reported *Niles' Weekly Register*, Jan. 10, 1824.
8. *Niles' Weekly Register*, Jan. 10, 1824.
9. "Report from the Select Committee on Manufactures, Commerce and Shipping . . . (1833)," in *British Parliamentary Papers, Industrial Revolution; Trade*, 2 (1833) (Shannon, 1968), p. 41 (hereafter cited as *BPP, Trade*); Edward Baines, *History of Cotton Manufactures in Great Britain* (London, 1835), p. 509; U.S. Congress, House, *Documents Relative to Manufactures in the U.S.* (Washington, D.C., 1833), 2:204, 220; Ware, *New England Cotton Manufacture*, pp. 41, 192; *BPP, Trade*, pp. 55, 170; *Niles' Weekly Register*, Nov. 24, 1827.
10. *BPP, Trade*, p. 150.
11. *Bulletin*, 5 Feb. 1880, in Borie Papers, Box 4, HSP.
12. Borie to E. Lords, Jan. 21, 1830, Borie and Laguerinne Letterbook, 1828–1831; Borie to Henry Clay, March 31, 1832, J. J. Borie and Sons Letter Book, 1829–1834, HSP (hereafter cited as Borie LB, 1829–34). On Keating, see Monsignor Eugene Murphy, *The Parish of St. John the Baptist, Manayunk: The First One Hundred Years, 1831–1931* (Philadelphia, n.d.), p. 172, HSP.
13. Wainwright, "Samuel Breck, 1827–1833," pp. 228, 230 n. 13; Hazard, *Register of Pennsylvania* 2(July–Jan. 1828):14–15. Borie and Laguerinne sent a great deal of their stock to Busto, Burckle, and Co. in Havana, who were heavily involved in the Spanish slave trade. Not only did they ship the coarse cloth for the African trade but muskets and gunpowder as well. Borie to Laguerinne and Bourell, Vera Cruz, April 12, 1830; Borie to Zimmerman, Frazer, and Co., Buenos Aires, Sept. 23, 1829, Borie LB, 1829–34; Borie to J. W. Zacharie and Co., Tampico, April 6, 1833; Borie to Busto, Burckle, and Co., Havana, April 20, 1833, Borie and Sons Letterbook, 1833–1834, HSP (hereafter cited as Borie LB, 1833–34).
14. *Pennsylvanian*, July 25, 1832; Laurie, *Working People*, p. 18; Charles Robson, *The Manufactories and Manufactures of Pennsylvania in the Nineteenth Century* (Philadelphia, 1875), pp. 411–12. See Scranton's treatment of Ripka's "accumulation strategy" in *Proprietary Capitalism*, pp. 142–53.
15. Borie to Busto, Burckle, and Co., Havana, Jan. 15, 1834, Borie LB, 1833–34; *Democratic Press*, Feb. 10, 1829.
16. Priscilla Ferguson Clement, "The Philadelphia Welfare Crisis of the 1820's," *PMHB* 105 (April 1981):151, 153, table 1.
17. "Report from the Select Committee on Emigration: 1826," *British Parliamentary Papers, Emigration*, 2 (1826–27) (Shannon, 1968), pp. 3–7, 189, 226, 223 (hereafter cited as *BPP, Emigration*); William J. Bromwell, *History of Immigration to the United States . . . 1855* (New York, 1856), pp. 146, 176; Duncan Bythell, *The Handloom Weavers: A Study in the English Cotton Industry during the Industrial Revolution* (Cambridge, 1969), p. 75; David Montgomery, "The Shuttle and the Cross: Weavers and Ar-

tisans in the Kensington Riots of 1844," *Journal of Social History* 5 (Summer 1972):417.
18. Mathew Carey, *Reflections on the Subject of Emigration from Europe, with a View of Settlement in the United States* (Philadelphia, 1826), p. 27. On returning immigrants, see also *Niles' Weekly Register*, June 3, 1826.
19. Clement, "Welfare Crisis," p. 158. In 1833, 244 impoverished Irish received financial assistance from the society. There were 498 in 1834 and 190 in 1835 (Hibernian Society for Relief of Emigrants from Ireland, Minutes, March 17, 1834, 1835, 1836, HSP). In 1835 women composed 48 percent of Irish immigrants to America (Adams, *Ireland and Irish Emigration*, p. 194; and see BBP, *Emigration*, pp. 226–27).
20. Clement, "Welfare Crisis," p. 163; *Niles' Weekly Register*, Feb. 21, 1824; Mathew Carey, *Female Wages and Female Oppression*, vol. 2, July 3, 1835.
21. J. S. Skinner to Mathew Carey, Aug. 18, 1833, Edward Carey Gardiner Collection, Carey Set, Box 87, HSP; Mathew Carey, *An Appeal to the Wealthy of the Land, Number 9* (Philadelphia, 1833), p. 26; Mathew Carey, "Essay on the Public Charities of Philadelphia," in *Mechanics' Free Press*, Feb. 7, 1829.
22. BPP, *Trade*, p. 44. The age and gender breakdown of Philadelphia mills in the mid 1830s was similar to cotton mills in southern Massachusetts. Ripka, for example, employed 79 men among his 413 workers in 1837 (PJS, 2:358, and passim; Prude, *Industrial Order*, p. 86).
23. A number of young single women also appeared in the Whitaker Accounts, Day Book, 1819–1823, Daily Ledger, Records of William Whitaker and Sons, Acc. 1471, Box 48, Eleutherian Mills–Hagley Foundation Library. Anthony Wallace describes the dominance of the family employment pattern in Rockdale through most of the century *Rockdale*, pp. 172, 177–78.
24. Harry Silcox, "Delay and Neglect: Negro Public Education in Antebellum Philadelphia, 1800–1860," *PMHB* 97 (Oct. 1973):449.
25. "The employment offered to children in the manufacturing establishments" was given as the reason for the high proportion of children in Manayunk's population. See Hazard, *Register of Pennsylvania* 2 (July–Jan. 1828):159.
26. *PJS*, 2:314, 315; *Public Ledger*, July 19, 1839.
27. *PJS*, 2:291.
28. Reuben Haines to Jane Haines, Oct. 17, 1830, Haines Correspondence, Wyck House, Germantown, Pa.
29. Kirk Boott, a Lowell mill owner, wrote Mathew Carey in 1827 that there were no foreigners in the mills except in the print works (White, *Samuel Slater*, p. 254). Dublin states that in 1836 the Hamilton Company of Lowell reported that only 4 percent of its labor force were foreign born (*Women at Work*, p. 26; see also Wallace, *Rockdale*, p. 65). The Pawtucket, Rhode Island mills, like Philadelphia's, drew on a labor pool of poor families (Kulik, "Pawtucket Village," p. 13). Prude's study of southern New England rural mills confirms that immigrants did not make up part of the operative labor force until mid-century (*Industrial Order*, pp. 87, 220–21).
30. BPP, *Trade*, p. 149; *PJS*, 2:325.
31. See testimony of Charles Hagner in *PJS*, 2:324.

32. Thompson, *The English Working Class*, pp. 248, 360. Thompson's statement is supported by political economist Andrew Ure in his 1835 treatise. See *The Philosophy of Manufactures; or, An Exposition of the Scientific, Moral, and Commercial Economy of the Factory System of Great Britain* (London, 1835; rpt. New York, 1967), pp. 20–21, 23; Dobb, *Development of Capitalism*, pp. 258–59; BBP, *Trade*, p. 168.
33. The throstles could not spin yarn above number 18. The mule was an intermittent spinning machine in which the stretching and twisting were simultaneous but the winding onto the bobbin was a separate step. The throstle stretched, twisted, and wound in one continuous operation. Thus the labor, time, and strength of a mule spinner was not required to reverse the spindles, wind on the yarn, and push back the carriage (David J. Jeremy, "Innovation in American Textile Technology during the Early Nineteenth Century," *Technology and Culture* 14 [1973]:47–48).
34. Quoted in Paul F. McGouldrick, *New England Textiles in the Nineteenth Century* (Cambridge, Mass., 1968), pp. 35–36.
35. *Niles' Weekly Register*, Oct. 23, 1830; Clark, *Manufactures in the U.S.*, p. 426, 532; Henry Carey, *Essay on the Rate of Wages: With an Examination of the Causes of the Differences in the Condition of the Laboring Population throughout the World* (Philadelphia, 1835), p. 79; Montgomery, *Practical Detail of Cotton Manufactures*, 2:204, 207, 220–21. The decrease in the price difference between a pound of cotton and a pound of yarn indicated the increasing output of spinning machinery. In 1815 the price differential was 27 cents; in 1830, 15 cents; and in 1841, 10 cents (Clark, *Manufactures in the U.S.*, p. 376).
36. Borie and Laguerinne to Mark A. Collet, Feb. 2, 1831, Borie and Laguerinne Letter Book, 1828–1831, HSP (hereafter cited as Borie LB, 1828–31).
37. Borie and Laguerinne to Goodwin, Rogers, and Co., Jan. 25, Feb. 2, March 22, April 5, Sept. 21, 1831, Borie LB, 1828–31; Borie and Laguerinne to Mark A. Collet, Feb. 2, 1831, Borie LB, 1828–31. The strike will be discussed in chapter 6.
38. Wallace, *Rockdale*, pp. 193, 381.
39. Baines, *Cotton Manufacture*, p. 495. On relative wage rates, see U.S. Congress, House, *Documents Relative to Manufactures*, 2:215; Hazard, *Register of Pennsylvania* 1 (Jan. 1828):28. See testimony before Senate investigative committee, PJS, 2; BBP, *Trade*, pp. 43, 166. The Philadelphia manufacturer James Kempton comments on the productivity of American powerlooms in his testimony before Parliament (BBP, *Trade*, pp. 168, 171). Samuel Slater shifted to the powerloom in the late 1820s for the same reasons Philadelphia's textile capitalists did (see Prude, *Industrial Order*, p. 123, 142).
40. Hazard, *Register of Pennsylvania* 1 (Jan.–July, 1828):28; 2 (July–Jan., 1828):14; Montgomery, "Shuttle and Cross," p. 417; Sullivan, *Industrial Worker*, p. 39; BBP, *Trade*, pp. 167–68; *Germantown Telegraph*, Jan. 1, 1834.
41. J. J. Borie and Son to James C. Kempton, April 26, July 30, 1834, Borie LB, 1833–34; Clark, *Manufactures in the U.S.*, p. 386; Ware, *New England Cotton Manufacture*, p. 94; Freedly, *Philadelphia and Its Manufactures*, p. 301; U.S. Congress, House, *Documents Relative to Manufactures*, 2:200–201. Borie intended to establish a calico printing manufactory to

increase profits. (J. J. Borie to Adolphus Borie, Dec. 30, 1825; Feb. 13, Oct. 12, 1826, Borie Papers, Box 3, F-4, HSP).
42. One can follow the state of the trades through *Niles' Weekly Register*, Aug. 25, Sept. 15, 1821; April 7, 1826; June 3, June 8, 1826; June 27, July 4, 1829; see also Hazard, *Register of Pennsylvania* 1 (Jan.–July, 1828):28.
43. H. Carey, *Essay on Wages*, p. 72.
44. Wallace, *Rockdale*, pp. 327–28.
45. *PJS*, pp. 324, 325.
46. My account of the factory system in Manayunk does not conform with Anthony Wallace's portrayal of the industrial village of Rockdale, some twenty-five miles southwest of Philadelphia. In this mill town, according to Wallace, men, machines, and nature operated in "harmonious combination" (*Rockdale*, p. 13).
47. Reuben Haines to Jane Haines, Oct. 17, 1830, Haines Correspondence.
48. Baines, *Cotton Manufactures*, p. 478; *BPP, Trade*, pp. 156, 169, 244, 658; U.S. Congress, House, *Documents Relative to Manufactures*, 2:221, 209. Prude, *Industrial Order*, pp. 130–31, 220.
49. *Pennsylvanian*, Aug. 28, 1833.
50. *PJS*, 1:322–23.
51. John Thornily, who had worked in the mills of Delaware County, cited "the want of uniformity in the working hours, at the different mills" as the "chief evil" of the factory system (*PJS*, 2:291).
52. Ibid., 2:286 and passim.
53. Ibid., 2:282, 294.
54. Ibid., 2:327.
55. Ibid., 2:292.
56. Makin testified that the piecers in Manchester, who tended two mules, walked twenty-five miles per day (ibid., 2:288, and see p. 286).
57. Ibid., 2:294.
58. Baines, *Cotton Manufactures*, p. 463; *PJS*, 2:293, 317; and see Sullivan, *Industrial Worker*, p. 317.
59. *PJS*, 2:318, 326.
60. See, for example, comment of Thomas Mosely, who had entered a Yorkshire factory at nine, immigrated, and had been working in Joseph Ripka's factory for nine years (ibid., 2:312–20).
61. *Pennsylvanian*, Aug. 28, 1833. Prude describes the labor of children in Dudley and Oxford cotton mills of southern Massachusetts as less harsh (*Industrial Order*, p. 128).
62. Montgomery, *Practical Detail of Cotton Manufactures*, pp. 97–98. It is difficult to draw comparisons of Philadelphia mill wage rates in the early and mid-1830s with those of Rockdale because Wallace's sources on specific rates are ambiguous. Wallace argues that the Rockdale owners paid "relatively high wages," but his evidence does not clearly confirm that claim. He has evidence on wage rates in the early 1830s from the Parkmount Mill Day Book, 1832–35. When he uses this source to aggregate family or individual annual incomes for 1832, weekly earnings are at Philadelphia levels. Rockdale's female throstle attendant wage rates, as given by Wallace, were comparable to Philadelphia's, but the date of his evidence is unclear (*Rockdale*, pp. 69, 172–78, 60, 140). In 1827, Kirk Boott of Lowell told Mathew Carey that the average daily wage in Lowell was 50 cents. Also compare the following discussion on Philadelphia wage

rates to Lowell's in White, *Samuel Slater*, p. 254. For British mill wage rates, see *BPP, Trade*, pp. 43, 70, 168, 169, 172, 83, 608, 647, and Prude, *Industrial Order*, p. 90. Ware indicates that rates in some Rhode Island mills were as low as the Philadelphia mills (*New England Cotton Manufacture*, p. 238).

63. *PJS*, 2:315.
64. This was the claim of the Working People of Manayunk in the *Pennsylvanian* of August 28, 1833. The Lancashire rates are taken from the English Factory Commissioners' Report of 1833 in Baines, *Cotton Manufactures*, p. 437. On Philadelphia rates, see *PJS*, vol. 2 passim, and Sullivan, *Industrial Worker*, pp. 41, 47. See also Montgomery, *Practical Detail of Cotton Manufactures*, p. 133; Dublin, *Women at Work*, p. 66; Ware, *New England Cotton Manufacture*, p. 239. The female operatives in the Whitaker factory in Philadelphia averaged three dollars per week because of the unusually high wage rates that female carding room attendants were paid (U.S. Congress, House, *Documents Relative to Manufactures*, 2:204).
65. Reuben Haines to Samuel Batchelder, Lowell, Dec. 8, 1830, Haines Correspondence; Ware, *New England Cotton Manufacture*, p. 230; Sullivan, *Industrial Worker*, p. 33.
66. *PJS*, 2:296–97. This pattern also existed in the Rockdale mills (Wallace, *Rockdale*, p. 179; Sullivan, *Industrial Worker*, p. 48).
67. See, for example, the working days lost at Henry Whitaker's mill between March 1838 and August 1839 in Henry Whitaker's Time Book, 1838, William Whitaker and Sons, Acc. 1471, Box 133. Eleutherian Mills–Hagley Foundation Library.
68. Borie to S. Lewis, President of Schuylkill Navigation Co., March 1, 1841, Borie Letter Book, Chief Justice Gibson, Paperbook ... 1842, p. 41, Pennsylvania Transportation, vol. 2, HSP.
69. Wallace's assertion is central to his assumption that the mobility of the mill workers was "an indication of financial ability to escape from unfavorable circumstances" (*Rockdale*, pp. 59, 63, 423). Mobility of Philadelphia mill workers and its relationship to class consciousness will be discussed in the following chapters.
70. Quoted in Pollard, *Modern Management*, pp. 184–85. Compare, for example, General Rules of the Schuylkill Factory, handbill, n.d., Borie Papers, Box 10, F-6, and "General Rules of the Silesia Factory" (Ripka's mill) in the *Germantown Telegraph*, Nov. 6, 1833.
71. *PJS*, 2:362, 280.
72. Ibid., 1:325; *Germantown Telegraph*, Oct. 30, 1833.
73. First quotation is from *PJS*, 2:296. Hagner's is in ibid., p. 326. Third quotation is in Memorandum of agreement made March 30, 1836 ..., Borie Papers, Box 1, F-8. Two sources indicate that operatives held the boss spinner in contempt for his role. See *Germantown Telegraph*, Oct. 30, 1830; *Roxborough News*, Nov. 30, 1927, in "Newspaper Clippings from the *East Falls Herald* and the *Roxborough News*," Scrapbook, HSP.
74. *PJS*, 2:32, 296, 297, 326; Dublin, *Women at Work*, p. 60.
75. *PJS*, 2:297.
76. One indicator of the relative labor scarcity in rural New England is that the Lowell and Rhode Island firms gave one-year contracts to workers and Philadelphia mills did not (Dublin, *Women at Work*, p. 60; White, *Samuel Slater*, p. 131).

77. Stanly Lebergott, *Manpower in Economic Growth: The American Record since 1800* (New York, 1964), pp. 127–28. Prude also describes a lenience among mill owners in enforcing rules because of the "tight labor market" (*Industrial Order*, pp. 115, 119, 155–56; see also Jeremy, *Transatlantic Industrial Revolution*, p. 253).

CHAPTER IV ECONOMY AND SOCIETY IN ROXBOROUGH AND MANAYUNK

1. Charles Byron Kuhlman, *The Development of the Flour-Milling Industry in the United States* (Cambridge, Mass., 1929), p. 29; Steven W. Fletcher, *Pennsylvania Agriculture and Country Life* (Harrisburg, 1950), p. 283. See Percy W. Bidwell and John I. Falconer, *History of Agriculture in the Northern United States, 1620–1860* (Washington, D.C., 1925), table 19, p. 136; table 66, p. 493; and pp. 134, 137, 139.
2. Robert Proud, *The History of Pennsylvania in North America*, (Philadelphia, 1798), 2:255; Rochefoucauld-Liancourt, *Travels*, 1:12. Horatio Gates Jones recalled there were eight flour mills in 1799 (see *Manayunk Star*, April 9, 1859).
3. Horatio Gates Jones, "Early History of Roxborough and Manayunk," 1855, Historical Society of Pennsylvania, Philadelphia; Hagner, *Early History*, p. 54.
4. See Jones in *Manayunk Star*, May 7, 1859. A merchant miller bought wheat and sold the processed flour. As a commercial manufacturer, he is distinguished from a custom miller who operated a service industry for the local farmers and received 10 percent of the product.
5. Charles Harper Smith, *The Livezey Family: A Genealogical and Historical Record* (Philadelphia, 1934), p. 41; Rochefoucauld-Liancourt, *Travels*, 1:4. Most of the roads built between 1730 and 1790 were mill roads petitioned by the Wissahickon millers who desired to connect their mill seats with the Ridge Road into Philadelphia. See Court of Quarter Sessions (hereafter cited as CQS), Road Dockets, Index, 1729–1790, Archives of the City and County of Philadelphia; Jones, "Early History," pp. 19, 29, 30; Jones in *Manayunk Star*, May 7, 1859; Douglas MacFarlan, "The Wissahickon Mills," 1947, HSP, 2:12, 46.
6. Oliver Evans, *The Young Mill-Wright and Miller's Guide* (Philadelphia, 1807), pp. 201–2, 238.
7. Kuhlman, *Flour-Milling Industry*, p. 100.
8. See "A List of the Subscribers' Names" in Evans, *Miller's Guide*; Rochefoucauld-Liancourt, *Travels*, 1:4, 7, 15. Samuel Wagner of Charlestown wrote Stephan Girard that Robeson had one of the two best brands of flour in the nation (Samuel Wagner to Stephen Girard, Jan. 1, 1817, Stephen Girard Papers, American Philosophical Society, Philadelphia).
9. Kuhlman, *Flour-Milling Industry*, p. 100. Only one grist mill was built on the Wissahickon after 1779 (Bishop, *History of American Manufacturers*, 1:142). In 1791 Peter Robeson was the only miller with real property valued in the highest wealth-holding category. In 1800 all but one of the eight mill owners moved into this category (Philadelphia County, Tax Assessors Ledgers, Roxborough, 1791, 1800, 1809; Tax Duplicate, Roxborough, 1819, Archives of the City and County of Philadelphia, Philadelphia [hereafter cited as Tax List]). The County Tax Duplicate of 1819 varies from the assessment ledgers in that it does not enumerate property but specifies the total tax. (See table at top of page 195.)

WEIGHTED PROPERTIED WEALTH CATEGORIES IN DOLLARS, 1791–1819

Year	Index[a]	$1–499	Base Categories $500–999	$1,000–2,499	$2,500+
1791	92	1–544	555–1,089	1,090–2,724	2,725+
1800	135	1–369	370–739	740–1,849	1,850+
1809	125	1–399	400–799	800–1,999	2,000+
1819	128	1–389	390–779	780–1,949	1,950+

Source: "Geometric Indices of Average Monthly Wholesale Prices in Philadelphia for Farm Derivatives," in Anne Bezanson, Robert D. Gray, and Miriam Hussey, *Wholesale Prices in Philadelphia, 1784–1861* (Philadelphia, 1948), chart 9, pp. 98–99; chart 10, pp. 158–59; chart 12, pp. 232–33.
[a]Base index is 1821–25.

10. Jones, "Early History," pp. 5, 8–9. Joseph Starne Miles lists and maps out the titles and tracts of the early settlers in *A Historical Sketch of Roxborough, Manayunk, Wissahickon* (Philadelphia, 1940), pp. 77, 78, and insert opposite p. 75.
11. Horatio Gates Jones, "Sketch of the Rittenhouse Paper Mill, 1863," Item 328, HSP, pp. 6–7. MacFarlan, "Wissahickon Mills," 1:9–24, 27; Christian Lehman, "Draught of Henry Rittenhouse Mill Road . . . from his Mill Dam, 1764," Christian Lehman Papers, Box 1, Roxborough Folder, HSP; Helen C. Perkins, Wissahickon Scrapbook, vol. 1, 1900–1919, HSP. In 1779, four different Rittenhouses owned three and a half of the eight flour mills in Roxborough (Jones in *Manayunk Star*, April 9, 1859; Tax List, 1791, 1800, 1809).
12. Tax List, 1791, 1800, 1809.
13. Robeson owned the most valuable mill and more real property than any other miller in 1791, 1800, 1809, and 1819. See Tax List, 1791, 1800, 1809, 1819; Miles, *Historical Sketch*, p. 77; Smith, *Livezey Family*, p. 79; Rochefoucauld-Liancourt, *Travels*, 1:6; MacFarlan, Work Sheets, Historical Society of Montgomery County; MacFarlan, "Wissahickon Mills," 1:84.
14. Smith, *Livezey Family*, pp. 38, 41, 42; Jones in *Manayunk Star*, April 9, 1859; MacFarlane, "Wissahickon Mills," 2:17.
15. Smith, *Livezey Family*, pp. 38, 48, 77–78; Miles, *Historical Sketch*, p. 66. Tax List, 1791, 1800, 1809.
16. SM, 1820. On the course of the flour export trade see Bidwell and Falconer, *History of Agriculture*, table 66, p. 493; Fletcher, *Pennsylvania Agriculture*, p. 294.
17. Charles R. Barker, "The Haverford-and-Merion Road to Philadelphia: A Walk Over an Old Trail," *PMHB* 58 (1934):252–53; Horatio Gates Jones, *The Levering Family* (Philadelphia, 1858), p. 103; Horace H. Platt and William Lawton, *Freemasonry in Roxborough: The History of Roxborough Lodge No. 135 and Accepted Masons, 1813–1913* (Philadelphia, 1913), p. 118; *Pennsylvania Archives*, 3d ser., 16:24–32.
18. In CQS, 1791, Roxborough, Box 304, packet 15–241.
19. United States, Direct Tax of 1798, Roxborough; Tax List, 1791, 1800, 1809. No other resident except for Robeson came close to owning as much land in Roxborough as Nathan Levering.
20. In Dudley and Oxford, Massachusetts, the rural townships Jonathan Prude

has studied, the propertyless rate in the early 1800s was 30 percent and the economic elite controlled just 30 percent of assessed wealth (*Industrial Order*, pp. 6, 12).
21. Laurie, *Working People*, p. 9.
22. For 1781 valuation and tax figures for Roxborough Township see *Pennsylvania Archives*, 3d ser., 1897, 16:29–33; see also James T. Lemon, "Household Consumption in Eighteenth-Century America and Its Relationship to Production and Trade," *Agricultural History* 41 (1967):68.
23. Rochefoucauld-Liancourt, *Travels*, 1:3. Roberta Balstad Miller has shown how the building of the Erie Canal drove up the price of farm land in Onondaga County, New York, and thus caused a decline in farm size. See *City and Hinterland: A Case Study of Urban Growth and Regional Development* (Westport, Conn., 1978), p. 96.
24. Rochefoucauld-Liancourt, *Travels*, 1:6, 13; Joshua Gilpin, "Journal of a Tour from Philadelphia through the Western Counties of Pennsylvania in the Months of September and October, 1809," *PMHB* 50 (1926):67.
25. Papermakers made up a new sector of propertyless taxpayers following the embargo (Tax List, 1809, 1819; *Wissahickon Paper Mill, 1731–1834*, pamphlet, HSP; SM, 1820; U.S. Direct Tax of 1798, Roxborough; Tax List, 1791, 1800).
26. Twenty-one of the thirty-eight who composed the 1791 cohort left. The sample is inclusive of all skilled and unskilled taxables without real property and with a per-head tax. It is therefore biased to the extent that young married men in the same position are not considered (Tax List, 1791, 1800, 1809, 1819).
27. David Dauer, "Colonial Coopers: Rural Laborers in Philadelphia's Commercial Economy" (manuscript, 1980), pp. 40, 45. Given that labor was expensive in relation to materials, dividing the production into discrete steps and using the best tools would allow great gains in productivity. A cooper employed by Roxborough's Lawrence Miller, for example, may have performed only the task of hewing staves or nailing the hoops to the casks. Miller, who owned the lumber, workspace, and tools, accumulated the increased surplus value created by the collection of, but more important, the division of labor.
28. For an analysis of the relationship between commercial milling and the growth of the cooper's trade in colonial Philadelphia's hinterland, see Dauer, "Colonial Coopers." See also Direct Tax of 1798, Roxborough; Tax List, 1791, 1800. The proportion of coopers in Roxborough at the turn of the century was relatively high. The city of Philadelphia had 64 coopers out of 6,818 taxables in 1800. Roxborough had 34 coopers in a population of 202 taxpayers. See Richard G. Miller, "Gentry and Entrepreneurs: A Socioeconomic Analysis of Philadelphia in the 1790s," *Rocky Mountain Social Science Journal* 12 (January 1975):77–79.
29. Tax List, 1791, 1800, 1809, 1819; County of Philadelphia Tax, Roxborough, 1783, in *Pennsylvania Archives*, 3d ser., 16:650–54. The shop-owning coopers represented one-sixth of the occupation. The large Wissahickon millers also owned cooper shops.
30. Adams, "Wage Rates," pp. 415, 406. On the relative vulnerability of coopers to downturns in the export market, see Dauer, "Colonial Coopers," pp. 48–49. The experience of William Sanders was a telling if exceptional example of the unequal effects of the growth of production in the craft.

William Sanders's father was a subsistence farmer in Roxborough at the time of the Revolution. William took up the trade of cooper. Although propertyless in 1791, he, like most coopers, in the following decade accumulated a small amount of property. Sometime between 1800 and 1809 Sanders lost his few acres. In 1809 he appeared on the tax list as a propertyless laborer. In the next decade he left the community having failed at his chosen craft.

31. Stephanie Wolf, *The Urban Village: Population, Community, and Family Structure in Germantown, Pennsylvania, 1683–1800* (Princeton, 1976), p. 190.
32. *Democratic Press*, Sept. 24, 1827.
33. The population figures for Roxborough are in John Daly and Allan Weinberg, "Genealogy of Philadelphia County Subdivision," n.d., City and County Archives of Philadelphia, Philadelphia. Population figures for Manayunk are from Hagner, *Early History*, pp. 79–80; *Germantown Telegraph*, Dec. 1, 1830; *Manayunk Courier*, Jan. 15, 1848.
34. *Democratic Press*, Sept. 24, 1827; *Germantown Telegraph*, Jan. 1, 1834, Dec. 1, 1830. In 1836 Charles Hagner counted 541 dwellings in Manayunk (*Early History*, pp. 79–80).
35. Isaac Baird's statement is in the *Norristown Herald and Weekly Advertizer*, Dec. 17, 1825. Hagner's statement is in Samuel Hazard, *Register of Pennsylvania* 1 (Jan.–July, 1828):159.
36. Hazard, *Register of Pennsylvania*, 2 (July–Jan., 1828):14.
37. *Pennsylvanian*, July 25, 1832; Hagner, *Early History*, pp. 79–80; *Germantown Telegraph*, Jan. 1, 1834. See testimony of Charles Hagner in *PJS*, 2:322.
38. Compare Roxborough's enumeration lists of 1828 and 1835 to the 1819 tax assessor's duplicate. Philadelphia County, Tax Duplicate, Roxborough, 1819; Tax List, 1828, 1835; *Pennsylvanian*, July 25, 1832.
39. Hazard, *Register of Pennsylvania* 2 (July–Jan. 1828):14.
40. "New buildings are rising—the streets are improving—the road from Flat Rock Bridge to Ridge Turnpike will be McAdamized," a Manayunk resident boasted in 1829 (*Poulson's American Daily Advertizer*, Jan. 24, 1829). By 1833 the Schuylkill was bridged at Green Lane in the middle of Manayunk, "nearer the center of activities" than the bridge at Flat Rock dam (Miles, *Historical Sketch*, p. 89; Jones, "Early History," p. 27).
41. Young is quoted in Hazard, *Register of Pennsylvania* 2 (July–Jan. 1828):15; Hagner is quoted in *PJS*, 2:320.
42. Of 671 Manayunk taxables, 572 were propertyless in 1840. Of 179 textile workers identified in the 1840 tax list for Manayunk, all but eight were propertyless (Philadelphia County, Manayunk, Tax, 1840; Pennsylvania State Tax, Roxborough, 1841 [hereafter cited as State Tax 1841], Archives of the City and County of Philadelphia).
43. Ibid.; United States, Schedule of Population, Sixth Census of the United States, 1840, Roxborough, National Archives, Washington, D.C. (hereafter cited as Census, 1840). Scranton found that the propertyless rate among "identifiable" male textile workers in 1850 was 96 percent, (Scranton, *Proprietary Capitalism*, p. 258).
44. *Germantown Telegraph*, Jan. 16, 1833; Horatio Gates Jones, "Historic Notes of Olden Time in Roxborough and Manayunk," p. 52, HSP (mounted articles from the *Manayunk Star*, 1859).

45. Jones, "Early History," p. 29; George J. Kennedy, *Roxborough, Wissahickon, and Manayunk in 1891* (Wissahickon, 1891), pp. 99–100; Hagner, *Early History*, p. 71; Murphy, *St. John the Baptist*, p. 416; Wainwright, "Samuel Breck, 1827–1833," p. 230; J. J. Borie to John Levering, Feb. 9, 1832, Borie LB, 1829–34.
46. *Bulletin*, Feb. 5, 1880, in Borie Papers, Box 4, HSP; Borie to Henry Clay, March 3, 1832, Borie LB, 1829–34; Wainwright, "Samuel Breck, 1827–33," p. 228, 230 n. 13; Hazard, *Register of Pennsylvania* 2 (July–Jan. 1828):14–15. On Lowell mill owners see Robert F. Dalzell, "The Rise of the Waltham-Lowell System and Some Thoughts on the Political Economy of Modernization in Ante-Bellum Massachusetts," *Perspectives in American History* 9 (1975):232–33, 247, 249, 252, 263. See also Mildred Goshaw, "Material Relating to Mills and Mill Owners of Manayunk in the Nineteenth Century" (manuscript, 1972), pp. 67, 73–74, Roxborough Free Library, Roxborough; *Pennsylvanian*, July 25, 1832. On the mixed backgrounds of Manayunk's textile capitalists in 1850 and 1860, see Scranton, *Proprietary Capitalism*, pp. 251–52.
47. Manayunk, Tax List, 1840; Roxborough, State Tax List, 1841.
48. Hagner, *Early History*, pp. 31–32, 68; Goshaw, "Mills and Mill Owners," p. 81; Manayunk, Tax List, 1840; *Pennsylvanian*, July 25, 1832.
49. *Poulson's Daily Advertizer*, Nov. 11, 1824. See will of Anthony Levering reprinted in Goshaw, "Mills and Mill Owners," p. 2. Levering sold the land to developer John Towers, who sold it to Ripka in 1828 (*Manayunk Courier*, March 11, 1848).
50. Hagner, *Early History*, p. 67; Pennsylvania State Tax, Roxborough, 1832, City and County Archives of Philadelphia, Philadelphia (hereafter cited as Tax List, 1832); Kennedy, *Roxborough*, pp. 8–9; John Levering, *Levering Family History and Genealogy* (Indianapolis, 1897), pp. 894–95; State Tax, 1841.
51. Indentures, 1836, Box I, F-2. Borie Papers, HSP. Robeson's mill was still considered one of the most prominent in the area in 1833 (Wainwright, "Samuel Breck, 1827–1833," p. 246).
52. Daniel K. Cassel, *History of the Rittenhouse Family* (Germantown, 1894), p. 19.
53. Hazard, *Register of Pennsylvania* 2 (July–Jan. 1828):15.
54. See Mildred Goshaw, "Gravestone Inscriptions in Graveyards of the Roxborough and Manayunk Areas," manuscript, Roxborough, 1967, HSP; Miles, *Historical Sketch*, p. 174. Murphy, *St. John the Baptist*, pp. 416, 43–44. Fifty-six of the burials at Saint John's were infants, two years old and younger. At least half had Irish surnames. See Burial Records reprinted in ibid., pp. 73–75. Philip Scranton's analysis of the 1850 census reveals that 57 percent of male workers were English- or Irish-born at midcentury (*Proprietary Capitalism*, p. 267, table 7.4).
55. Sources for the individual cases in this cohort as discussed below are as follows: Thornily: *PJS*, 2:290–92; Mosely: *PJS*, 2:315–16; Wilkinson: *PJS*, 3:319–20; Milligan: *PJS*, 2:310–12; Siddall: *PJS*, 2:312–13; Boulton: *PJS*, 2:318–19; Ogden: *PJS*, 2:296–99, *Pennsylvanian*, June 3, 1834; Neld: *PJS*, 2:293–94; Craig: *PJS*, 2:283–85, and see Tax List, 1835; Bowker: Kennedy, *Roxborough*, p. 87; Foster: Robson, *Manufacturers of Pennsylvania*, p. 132; Loughrey: Murphy, *St. John the Baptist*, p. 222; White: Platt and Lawton, *Freemasonry in Roxborough*, p. 131; Rudolph: Kennedy, *Rox-*

borough, p. 90; Gartside: Robson, *Manufacturers of Pennsylvania*, p. 91; Morris: Charlotte Erickson, *Invisible Immigrants: The Adaptation of English and Scottish Immigrants in Nineteenth-Century America* (Coral Gables, Fla., 1972), pp. 140–43.

56. The evidence I have found on Irish and British immigrants in Philadelphia's factory labor force challenges Bruce Laurie's conclusion that an Irish "aversion to modern work discipline" kept them away from Philadelphia factories and manufactories in the pre-1840 period (Laurie, *Working People*, pp. 28–29). On the immigrant population in Manayunk and its factories in the late antebellum period, see Scranton, *Proprietary Capitalism*, p. 253, 261–62.

57. Anthony Wallace found a similar out-migration rate for one of the nearby Rockdale mills. See *Rockdale*, pp. 63–64.

58. Power Loom Accounts, William Whitaker and Sons, Accounts, Acc. 1471, no. 131, Eleutherian Mills–Hagley Foundation Library. Thomas Dublin found that women in the Hamilton Company mills in July 1836 averaged fourteen months between arrival and departure (Dublin, *Women at Work*, pp. 184–85).

59. The experiences of William Shaw and Luke Neld, Philadelphia textile employees for whom there is no evidence indicating they were in Manayunk, confirm that mobile factory workers were generally on the move to other mills. Shaw labored in four different Rockdale and Philadelphia mills as a young mule spinner between 1828 and 1837. In 1837 Neld had accumulated thirty years' experience in mills in Manchester, Rockdale, and Philadelphia (*PJS*, 2:279–81, 293–94). Many of Manayunk's immigrant labor force came straight to the mill town from the docks in Philadelphia, and some became permanent residents. Four of the English-born textile operatives who testified before the factory investigative committee in 1837 had been employed by Joseph Ripka or James Kempton since they had immigrated up to five years before. English-born James White, as noted earlier, arrived in Manayunk in 1814 and was a spinner in the mills until 1848 (Platt and Lawton, *Freemasonry*, p. 131).

60. Wallace, *Rockdale*, p. 63.

61. Wallace, *Rockdale*, p. 423. Jonathan Prude has found a high mobility rate among the mill workers of southern Massachusetts throughout the antebellum period. He interprets "leaving" as a form of protest. See *Industrial Order*, p. 144, 146, 148–50, 227.

CHAPTER V THE INSTITUTIONS OF ORDER

1. Wolf, *Urban Village*, pp. 177–78.
2. My analysis of Roxborough's social relations draws much from Howard Newby's synthesis in "The Deferential Dialectic," *Comparative Studies in Society and History* 17 (April 1975):156.
3. Platt and Lawton, *Freemasonry*, p. 113; George Allen, "The Rittenhouse Paper Mill and Its Founder," *Mennonite Quarterly Review*, April 1942, p. 23 (offprint in the Eleutherian Mills–Hagley Foundation Library); Tax List, 1791. Martin Rittenhouse housed those who worked in his mill. U.S., Direct Tax of 1798, Roxborough; Tax List, 1800.
4. Quote is in Horatio G. Jones, "Historic Notes of Olden Time in Roxborough and Manayunk" (compilation of articles first published in the *Manayunk Star*, 1859, and reprinted in the *Manayunk Sentinel and Star*, 1880),

HSP, p. 56, see also pp. 42, 52. The large Wissahickon merchant millers—Robeson, Livezey, and the Rittenhouses—all served as managers of the poor house or directors of Roxborough's school in the late eighteenth and early nineteenth century.

5. *Manayunk Courier*, March 4, 1848. On Borie, see Borie to Hagner, Jan. 22, 1833, Hagner Collection, HSP; quotation on Ripka is in Hagner, *Early History*, p. 71; see also Jones, "Early History," p. 29; Kennedy, *Roxborough*, pp. 99–101; Manayunk Council Minutes, July 14, 1840, City and County Archives of Philadelphia; Boles, Penrose Pictorial Collection, no. 27, Falls, Roxborough, Manayunk, HSP.
6. Hagner, *Early History*, p. 86–88; Hazard, *Register of Pennsylvania* 1 (Jan.–July 1828):159–60.
7. *Manayunk Courier*, March 4, 1848; Hagner, *Early History*, p. 86. Wallace argues similarly that when the mill owners of Rockdale had to balance economic interest with child welfare or education, the latter was regularly accorded second place (*Rockdale*, pp. 332–34; *PJS*, 2:282, 359).
8. *PJS*, 2:314, 291, 298. The figures below are drawn from Hagner's report to the Peltz Committee (*PJS*, 2:325) and from the 1840 census.
9. William Shaw testified that children attended Sabbath school with great reluctance: "Many will not attend in consequence of the confinement of the week" (*PJS*, 2:281, 289, 292, 313, 325; *Manayunk Courier*, March 4, 1848).
10. Records of the German Reformed Congregation in Germantown 1753–1856, vol. 2; Records of St. Michael's Evangelical Church in Germantown, Pa., HSP; Jones, "Early History," p. 15; "Historic Notes," chap. 40.
11. Immediately upon formation of the church, Nathan, Abraham, and John Levering were appointed to oversee the building of the meetinghouse. Nathan, "a very prominent and public spirited citizen" as well as the wealthiest person in Roxborough, donated the ground lot for the meetinghouse and supervised its construction (Jones, "Historic Notes," chap. 48). The discussion that follows on the Roxborough Baptist Church is drawn from Roxborough Baptist Church Membership Register, 1789–1909, HSP (hereafter cited as RBC, Register) and from Minutes of Roxborough Baptist Meeting, 1789–1858, HSP (hereafter cited as RBC, Minutes).
12. Of the sixty-one individuals who were baptized or received by letter between 1790 and 1820, twenty-one can be linked to the tax lists. All but a few of these members were tradesmen or farmers with little or no property (RBC, Register; Tax List, 1791, 1800, 1809, 1819).
13. Jones, *Roxborough Baptist Church*, p. 31. Exclusion was the maximum penalty and occurred only eight times before 1820. Women were warned and excluded primarily for "dancing and frolicking." Men were warned and excluded primarily for drinking.
14. On Graham and Linn, see RBC, Minutes, Sept. 27, 1817; Jan. 22, 1820.
15. Ibid., March 25, 1815; Jones, *Roxborough Baptist Church*, p. 14.
16. RBC, Minutes, Feb. 21, 1818. A fire in 1830 burned the records of the weekly meeting for 1820–30, which makes it difficult to analyze the small revival and wave of exclusions (Jones, "Early History," pp. 15, 16–18).
17. Grueber was a close disciple of America's strict Methodist bishop, Francis Asbury. See Mt. Zion Methodist Episcopal Church, Manayunk, His-

torical Record, Roxborough United Methodist Church, Roxborough; Laurie, *Working People*, p. 37.
18. A Register of the Names of Members Belonging to the M.E. Church in Manayunk, Jan. 1837, Roxborough United Methodist Church, Roxborough.
19. Ibid.; Tax List, 1835. I examined the household composition of all female-headed households that contained a member or members who were employed in manufacturing (Census, 1840).
20. Methodist Historical Record.
21. Thompson, *The English Working Class*, p. 362.
22. Methodist Historical Record.
23. Laurie, *Working People*, pp. 46–48; Paul E. Johnson, *A Shopkeeper's Millennium: Society and Revivals in Rochester, New York, 1815–1837* (New York, 1978), pp. 119–28.
24. Goshaw, "Mill Owners," p. 74; Records of the Fourth Reformed Church, Manayunk, Philadelphia, 1831–1914, HSP; Tax Lists, 1828, 1835; State Tax, 1841; Goshaw, "Gravestone Inscriptions"; Miles, *Historical Sketch*, pp. 178, 186; *Germantown Telegraph*, Jan. 4, 1832; *Weekly Forecast*, Nov. 23, 1916, HSP.
25. Murphy, *St. John the Baptist*, pp. 62–70, 165, 412; Tax Lists, 1828, 1835.
26. For Barat's recollections, see Murphy, *St. John the Baptist*, pp. 47, 49.
27. Ibid., p. 44.
28. Hagner, *Early History*, pp. 91–92; *Norristown Herald and Weekly Advertizer*, Oct. 21, 1855; Miles, *Historical Sketch*, p. 180; Platt and Lawton, *Freemasonry*, pp. 120–23.
29. Jones, *Roxborough Baptist Church*, p. 15. See RBC, Register.
30. MacFarlan, "Wissahickon Mills," 2:3. See also U.S., Bureau of the Census, *First Census of the United States, 1790, Heads of Families, Pennsylvania* (Washington, D.C., 1908), pp. 207–8; Thomas H. Shoemaker, "A List of Inhabitants of Germantown and Chestnut Hill in 1809," *PMHB* 15 (1891): 447–80; Draught of Henry Rittenhouse Mill Road, 1764, Christian Lehman Papers, Box 1, Roxborough Folder, HSP; Wolf, *Urban Village*, pp. 75, 192.
31. "Orders of Court to Lay Out Henry Rittenhouse Mill Road Up to Roxborough Great Road," Sept. 7, 1767, "Christian Lehman Papers," Box 2, HSP.
32. CQS, Dec. 1791, Road Dockets, Roxborough. If citizens of the county desired a public road or improvement, they could submit a petition to the County Court of Quarter Sessions for consideration. The court would appoint a committee of three to review the case.
33. CQS, March 1792, Road Dockets, Roxborough.
34. Ibid.
35. CQS, Dec. 1791, March 1792, Road Dockets, Roxborough, Petitions. The Martin Rittenhouse Road was built despite the protestations of most of the inhabitants, and in 1794, John Levering, Michael Righter, and William White petitioned for a "Valuation of the grounds over which Rittenhouse Road passes." They were awarded a few pounds each in damages (CQS, 1794, Road Dockets, Roxborough; Platt and Lawton, *Freemasonry*, p. 113).
36. CQS, June 1811, Road Dockets, Roxborough.
37. CQS, Sept. 1811, Road Dockets, Roxborough.

38. CQS, Dec. 1811.
39. Jones, "Early History," p. 19.
40. Jones, "Roxborough and Manayunk," p. 73.
41. Hagner, *Early History*, pp. 85–86. The individual mill owners willingly forwarded their subscriptions.
42. CQS, Road Dockets, 1829, 50–759, 1831, 54–857.
43. CQS, March, June 1828, Roxborough; Sept. 1830, Falls of Schuylkill; Jan. 1833, Penn Township; March 1834, Sept. 1835, Roxborough; Jones, "Early History," p. 210.
44. Jones, "Early History," p. 27.
45. The road that the merchant miller proposed would cost between ten and fifteen thousand dollars (CQS, Road Petitions, 1833, Roxborough).

CHAPTER VI PRIMITIVE PROTEST, REPUBLICAN REFORM, AND RELIGIOUS REVIVAL

1. Thompson, *The English Working Class*, p. 191.
2. *Germantown Telegraph*, July 31, 1839; Robson, *Manufacturers of Pennsylvania*, p. 463. Scranton addresses the question of the dynamics of fires in the context of "capitalist exploitation and accumulation" (*Proprietary Capitalism*, pp. 79–80).
3. The armed men who met the weavers cannot be identified from available sources (Freedley, *Philadelphia and Its Manufacturers*, p. 301; Kennedy, *Roxborough*, p. 100.
4. *Public Ledger*, Sept. 26, 1842, quoted in Michael Feldberg, "Urbanization as a Cause of Violence: Philadelphia as a Test Case," in Allen F. Davis and Mark H. Haller, eds., *The Peoples of Philadelphia: A History of Ethnic Groups and Lower-Class Life, 1790–1940* (Philadelphia, 1973), p. 64.
5. Twenty-nine hundred weavers in Philadelphia went on strike in 1825 to obtain higher piece rates (*Niles' Weekly Register*, May 7, 1825). Scranton discusses the dynamics of this conflict in *Proprietary Capitalism*, pp. 124–25.
6. Niles reported the first story in *Niles' Weekly Register*, Aug. 16, 1828. On the second incident, see Michael Feldberg, "Urbanization as a Cause of Violence," p. 57.
7. Hammond and Hammond, *The Skilled Labourer*, pp. 126–27, 273; George Rudé, *The Crowd in History, 1730–1848* (New York, 1964), pp. 85–86; Thompson, *The English Working Class*, pp. 565–67; R. C. Kirby and A. E. Musson, *Voice of the People: John Doherty, 1798–1854: Trade Unionist, Radical, and Factory Reformer* (Manchester, 1975) p. 41. New York weavers also became machine breakers in the classic tradition of the Spitalfield weavers and Lancashire Luddites (Wilentz, *Chants Democratic*, pp. 170–71).
8. Laurie, *Working People*, p. 75. Louis H. Arky, "The Mechanics' Union of Trade Associations and the Formation of the Philadelphia Workingmen's Movement," *PMHB* 76 (April 1952):149; William Heighton, *An Address to the Members of Trades Societies and to the Working Classes Generally* (Philadelphia, 1827).
9. Kirby and Musson, *Voice of the People*, p. 158.
10. Foster, *Class Struggle*, pp. 18–19, 80–81. Ware, *New England Cotton Manufacture*, pp. 97–98.

11. Fifth Annual Report of the Union Benevolent Society, 1836, HSP; Sullivan, *Industrial Worker*, p. 155.
12. Sullivan, *Industrial Worker*, pp. 223–30.
13. Laurie, *Working People*, p. 85; Edward Pessen, *Most Uncommon Jacksonians: The Radical Leaders of the Early Labor Movement* (Albany, 1967), p. 51; Arky, "The Mechanics' Union of Trade Associations," pp. 158–59.
14. This challenges Anthony Wallace's argument that the spinners in Philadelphia did not organize as other trades and were not inspired by the union movement (*Rockdale*, p. 255). See also Sullivan, *Industrial Worker*, p. 101.
15. Hammond and Hammond, *The Skilled Labourer*, pp. 93, 99; Kirby and Musson, *Voice of the People*, pp. 13, 15, 18, 43, 147–48. See also G. D. M. Cole, *Attempts at General Union: A Study in British Trade Union History, 1818–1834* (1953; rpt. Westport, Conn., 1979), pp. 51, 57.
16. *Mechanics' Free Press*, Dec. 20, Dec. 27, 1828. The United Brotherhood Society of Journeymen Cordwainers was William Heighton's union and he urged them to support Manayunk's spinners. There is no evidence that they received organized support from Manayunk's tradesmen (Arky, "The Mechanics' Union of Trade Associations," pp. 158–59 n. 33; Sullivan, *Industrial Worker*, p. 146).
17. See law case published in Hazard, *Register of Pennyslvania* 3 (Jan.–July 1829):39.
18. See Kirby and Musson, *Voice of the People*, p. 13.
19. Joseph Rayback, *A History of American Labor* (1954; rpt. New York, 1966), pp. 56–57, 59; Kirby and Munson, *Voice of the People*, p. 33.
20. Hazard, *Register of Pennsylvania* 3 (Jan.–July 1829):39.
21. Sullivan, *Industrial Worker*, p. 146.
22. Borie to Mark Richards, Dec. 4, 1829, Borie LB, 1828–31.
23. Borie did share with Richards one tactic that could alleviate the short-term weaver shortage. By taking "girls from our carding room and other parts" of the factory and placing them in the weaving rooms with a one-year contract, Borie had relieved his deficiency in the powerloom room (ibid.).
24. Hazard, *Register of Pennsylvania* 3 (Jan.–July 1829):39.
25. Heighton, *An Address Delivered Before the Mechanics & Working Classes Generally, of the City and County of Philadelphia by the 'Unlettered Mechanic'* (Philadelphia, 1827), Library Company of Philadelphia.
26. See William Sullivan, "Did Labor Support Andrew Jackson?" *Political Science Quarterly* 62 (Dec. 1947):569–80; Arky, "The Mechanics' Union of Trade Associations," passim; Pessen, *Most Uncommon Jacksonians*. Sean Wilentz argues that a radical working men's movement in New York must be distinguished historically from the Workingmen's Party (*Chants Democratic*, pp. 212–13).
27. *Mechanics' Free Press*, June 14, June 28, 1828.
28. Ibid., Aug. 9, Oct. 11, 1828.
29. *Mechanics' Free Press*, Aug. 22, 1829. Roxborough's Freemasons, an organization of prosperous tradesmen, shopkeepers, and proprietors, were well represented in the Workingmen's Party (Platt and Lawton, *Freemasonry*, pp. 266–70).
30. *Mechanics' Free Press*, July 25, Aug. 22, 1829.

31. Joseph Clark, a storekeeper, was elected delegate to the county convention (ibid., Aug. 22, 1829).
32. *Mechanic's Free Press*, Oct. 17, Oct. 25, 1829, Aug. 14, 1830. Wilentz traces this process among New York's workingmen. See *Chants Democratic* pp. 208–10, and Sullivan, "Did Labor Support Jackson," p. 572.
33. F. Murphy to Horatio J. McClean, Feb. 2, 1828, MSA 398, Boston Public Library. Hazard, *Register of Pennsylvania* 2 (July–Jan. 1828):15.
34. Arthur M. Schlesinger, Jr., *The Age of Jackson* (Boston, 1945), pp. 142–43, 320. David Montgomery develops the point that nativism did not divide burgeoning working-class cohesiveness until after the 1837–42 depression. It seems plausible that both Catholic and Protestant immigrant workers would be part of the Democracy. See Montgomery, "Shuttle and Cross," pp. 411–39.
35. *Pennsylvanian*, Oct. 6, 1832. The first question refers to fellow Irish who fought in the Irish rebellion of 1798. Wilentz argues that the Jacksonian Democrats took up the language of the working men as a self-conscious political decision (*Chants Democratic*, p. 214).
36. *Germantown Telegraph*, Sept. 5, 1832, Sept. 26, 1832.
37. Thompson, *The English Working Class*, p. 189. Jonathan Prude discusses the tension with which the towns of Dudley and Oxford viewed the new population of operatives and mill owners. See Prude, *Industrial Order*, pp. 178–79 and chap. 6, passim.
38. *Germantown Telegraph*, Aug. 2, 1837. Hagner told the Senate's factory investigation committee the same thing (*PJS*, 2:323).
39. Records of the Fourth Reformed Church, Manayunk, Philadelphia, 1831–1914, HSP; RBC, Membership.
40. *Minutes of the Philadelphia Baptist Association*, 1833, HSP (hereafter cited as *PBA, Minutes*).
41. Quoted in Jones, *Roxborough Baptist Church*, p. 57.
42. Bruce Tucker, "Class and Culture in Recent Anglo-American Religious Historiography: A Review Essay," *Labour/LeTravailleur* 6 (Autumn 1980): 164. See also Johnson, *Shopkeeper's Millennium*; Mary Ryan, *Cradle of the Middle Class: The Family in Oneida County, New York, 1790–1865* (Cambridge, Mass., 1981); Wallace, *Rockdale*.
43. Miles had been a member of the Workingmen's Party in 1828 and 1832. Levering had joined the meetings of the anti-Jacksonians in the fall of 1832 and was nominated by them for tax assessor (RBC, Minutes).
44. Johnson, *Shopkeeper's Millennium*, p. 138.
45. Records of the Fourth Reformed Church, Manayunk, Philadelphia; Marion C. Bell, *Crusade in the Cities: Revivalism in Nineteenth-Century Philadelphia* (Lewisburg, Pa., 1977), pp. 61–62, 98; Miles, *Historical Sketch*, p. 186.
46. Eric Hobsbawm, *Primitive Rebels: Studies in Archaic Forms of Social Movement in the Nineteenth and Twentieth Centuries* (New York, 1959), p. 132.
47. The number of new Baptist and Presbyterian members among the population of mill operatives was relatively small. The fact that perhaps half of the operatives were Irish and German Catholics and communicants in Jerome Keating's congregation, however, makes the evangelical Protestant converts a sizable sector of the potential pool of unchurched mill workers (RBC, Minutes).

48. No male with the same surname as the five females appeared in the records of the township. A small number of married couples—one dozen—can be identified with certainty as taking part together in the evangelical conversions of Nichols's pastorate. Three of the husbands can be identified as laborers, one was a wheelwright, another a grocer. The wealthy mill owner, Enoch Levering, and his wife Sophia also came into their forefathers' church in this period (RBC, Minutes).
49. Hobsbawm, *Primitive Rebels*, pp. 130–31; Thompson, *The English Working Class*, pp. 379–82.
50. Mary Ryan has argued that middle-class women were attracted to the revivals in Utica because they were concerned about their own salvation and that of their children in an uncertain capitalist world (*Cradle of the Middle Class*, pp. 77, 79, 80).

CHAPTER VII THE MILL STRIKES AND THE NEW IDEOLOGY OF CLASS

1. Alexis de Tocqueville, *Democracy in America* (1835; rpt. New York, 1956), p. 220.
2. I draw here upon Anthony Gidden's definition of class consciousness. See *The Class Structure of Advanced Societies* (London, 1981), pp. 111–12.
3. See G. D. H. Cole, *Socialist Thought: The Forerunners, 1789–1850* (London, 1952), pp. 105–8, 111. Ronald Schultz argues that Philadelphia working-class thought was influenced by English radical economists in the early national period. However, he does not consider the transmission of those ideas by industrial immigrants. See Ronald Douglas Schultz, "Thoughts among the People: Popular Thought, Radical Politics, and the Making of Philadelphia's Working Class, 1765–1828" (Ph.D. diss., University of California, Los Angeles, 1985), p. 392 and chapter 6, passim.
4. *Germantown Telegraph*, Aug. 14, Aug. 21, 1833.
5. Clark, *History of Manufactures*, pp. 279–80.
6. *Germantown Telegraph*, Aug. 14, 1833.
7. Cole, *Attempts at General Union: A Study in British Trade Union History, 1818–1834* (London, 1953), p. 5 and see pp. 7, 15, 23. *Germantown Telegraph*, Aug. 14, 1833. Scranton has determined that over three quarters of the spinners over the age of fifty in Manayunk's mills in 1850 were English-born. They were one-fourth of the male textile labor force (*Proprietary Capitalism*, pp. 257–58).
8. On Mullin and McGlinchy, see Murphy, *St. John the Baptist*, pp. 43–44.
9. On Gilmore, see Sullivan, *Industrial Worker*, pp. 103–4; Laurie, *Working People*, pp. 114–15. On Small, see ibid., p. 111.
10. *Pennsylvanian*, Aug. 28, 1833.
11. Ibid.
12. Ibid.
13. Ibid.
14. Kirby and Musson, *Voice of the People*, p. 175.
15. See Hagner in *PJS*, 2:326.
16. Before the turn of the century, Lancashire female spinners were members of the Manchester spinner's society. See Ivy Pinchbeck, *Women Workers and the Industrial Revolution, 1750–1850* (London, 1930), pp. 308, 134.
17. Children mill operatives in nearby Norristown harassed a strike breaker so persistently during a dispute in 1828 that he brought suit against them. See Commons, *History of Labour*, 1:418–19.

18. Quoted in *Roxborough News,* Nov. 20, 1927, in A. C. Chadwick, "Newspaper Clippings from the *East Falls Herald* and the *Roxborough News,*" scrapbook, HSP.
19. See *Germantown Telegraph,* Sept. 18, Oct. 16, Oct. 23, Oct. 30, Nov. 6, Nov. 13, and Nov. 20, 1833.
20. Ibid., Aug. 21, 1833.
21. Ibid., Aug. 28, 1833. See Pessen, *Uncommon Jacksonians,* p. 119. "A Workingman" drew on the same body of ideas and convictions as John Doherty. See Kirby and Musson, *Voice of the People,* p. 322.
22. *Germantown Telegraph,* Sept. 18, 1833.
23. Ibid., Aug. 28, 1833. On the mechanic's ideology see Faler, *Mechanics and Manufacturers,* pp. 175–76.
24. Kirby and Musson, *Voice of the People,* p. 322. This was the basis for Robert Owen's proposal to overcome the laws of the competitive market. If the producers produced only for themselves through the cooperative system, "commercial cupidity and class rivalries" would end. The nonproducers would be compelled to join the system (ibid., p. 272). See also Cole, *Socialist Thought,* pp. 106–8; J. E. King, "Utopian or Scientific? A Reconsideration of the Ricardian Socialists," *History of Political Economy* 15, no. 3 (1983): 345–73.
25. *Germantown Telegraph,* Aug. 28, 1833. John Ferral stated that "The Observer" was hired by the mill owners to answer "A Workingman" (ibid., Oct. 30, 1833).
26. *Germantown Telegraph,* Nov. 27, 1833. On "The Observer," see also ibid., Nov. 6, 1833. On "The Perceiver," see ibid., Nov. 20, 1833.
27. Ibid., Oct. 30, Nov. 13, Nov. 27, 1833. On November 6, one reader complained that the opponents had "abused each other enough." A letter from each of the four correspondents remained unpublished (ibid., Dec. 4, 1833). Observer and his opponents were not exposing a controversy unfamiliar to the industrial community. Hundreds of laboring-class residents, galvanized by the influence of Manchester's "wild and extravagant theories," had participated in a riot surrounding the lecture visit of socialist Frances Wright D'Arrusmont to the Falls of Schuylkill in July 1829. For a colorful description of the event, see Phillip M. Hagner to George Hagner, July [1829?], Charles V. Hagner Society Collection, Case 20, Box 22, HSP.
28. These ideas figured importantly in the oppositional stance of organized labor in Lynn, Lowell, and New York. They are further defined in Faler, *Mechanics and Manufacturers,* pp. 185–87; Dublin, *Women at Work,* pp. 88–94; and Wilentz, *Chants Democratic.*
29. Schultz, "Thoughts among the People," pp. 392–96. On Lynn see Faler, *Mechanics and Manufacturers,* pp. 181–83. On New York see Wilentz, *Chants Democratic,* pp. 238–42.
30. *Pennsylvanian,* Aug. 22, Aug. 28, 1833; John R. Commons et al., eds., *Documentary History of American Industrial Society,* (1911; rpt. New York, 1958), 5, part 1: 325; *Germantown Telegraph,* Aug. 14, 1833.
31. Workingman wrote this on August 21 at the same time that the acting committee was drafting the statement for their meeting on the twenty-third (*Germantown Telegraph,* Aug. 28, 1833).
32. Ibid., Oct. 23, Oct. 30, 1833; *Pennsylvanian,* Nov. 11, 1833.
33. Sean Wilentz has analyzed the roots and nature of New York's trade unionist

critique of early industrial capitalism. He does not locate the origins of this "class conscious trade unionism" in industrial Britain, which I argue was the wellspring of oppositional thought of Manayunk and Philadelphia's factory trade unionists. Rather, he notes that close parallels existed between expressions of working-class consciousness in New York and abroad (*Chants Democratic*, p. 243, and see passim, pp. 232–248).

34. Cole, *Socialist Thought*, p. 121; Cole, *Attempts at General Union*, pp. 116, 120; Kirby and Musson, *Voice of the People*, pp. 196, 323.
35. Cole, *Socialist Thought*, p. 111; Kirby and Musson, *Voice of the People*, pp. 321, 322, 323. Wilentz charts the development after 1834 of New York's General Trades' Union from a moderate to radical position to what he calls a "distinctly American trade unionism" (*Chants Democratic*, pp. 238, 239–44).
36. On Ripka, see *PJS*, 2:337–38. On British labor movement, see Thompson, *The English Working Class*, pp. 774, 796; Kirby and Musson, *Voice of the People*, p. 153; and Foster, *Class Struggle*, p. 108.
37. On the TUCCP, see Laurie, *Working People*, pp. 86–87; *Germantown Telegraph*, Nov. 6, Nov. 27, Dec. 24, 1837.
38. Laurie, *Working People*, pp. 86–87; Commons et al., *History of Labour*, 1:313; Commons et al., *Documentary History*, 5, part 1: 192.
39. *Pennsylvanian*, Dec. 28, 1833.
40. Ibid., Dec. 24, 1833. *Germantown Telegraph*, Nov. 6, 1833.
41. While Anthony Wallace has rightly observed that Philadelphia mill workers did not have a formal trade organization that endured from strike to strike, as did the skilled crafts, it is clear that an oppositional cadre of Manayunk mill workers had not dissolved during the winter (*Rockdale*, p. 367). A year-long series of strikes by cotton spinners countered a series of announced wage reductions in Manchester in 1826 (Kirby and Musson, *Voice of the People*, p. 43).
42. *Germantown Telegraph*, March 19, 1834.
43. The assembled strikers resolved to endeavor to find other jobs rather than work "from daylight until 8 o'clock in the evening, for that which will not procure us the necessary articles to sustain our miserable existence." The textile operative made the point that the room bosses in the factory had betrayed them in not counseling them "in the proper measures to withstand the reductions." They welcomed the advice and support of the sympathetic tradesmen in Manayunk as well as in Philadelphia. See *Germantown Telegraph*, March 19, 1834; Kirby and Musson, *Voice of the People*, p. 161.
44. Sullivan, *Industrial Worker*, p. 147. See handbill titled "Caution . . ." in Borie Papers, Box 9, F-1, HSP.
45. Ibid. See F. P. Baras to Messr. Borie and Laguerinne, April 29, 1834, Borie Papers, Box 9, F-1, HSP.
46. Barat to Borie and Laguerinne, April 29, 1834, Borie Papers, Box 9, F-1.
47. Ibid.
48. Joseph Clark, storekeeper, Episcopal vestryman, and former member of the Workingmen's Party, chaired the meeting of May 9.
49. *Germantown Telegraph*, May 15, 1834.
50. I can find no specific evidence of the settlement of the strike. William Sullivan claims that the operatives went back to work with an increase. See *Industrial Worker*, p. 148.

51. Quoted in Commons et al., *Documentary History*, 6, part 2: 223–24; see also pp. 196, 229.
52. See "Plain Truth" in the *Germantown Telegraph*, Aug. 2, 1837. Portions appear as well in Hagner's testimony before the Peltz Committee (*PJS*, 2:320–28). Hagner, as a true Jacksonian, blamed the cycles of expansion and contraction of the currency as the reason manufacturers were forced to reduce wages, which in turn caused "the frequent difficulties between [employer and employee]."

CHAPTER VIII THE POLITICS OF CLASS CONFLICT IN EARLY INDUSTRIAL SOCIETY

1. Prude, *Industrial Order*, p. 106.
2. The *Pennsylvanian* was referring to the national bank's policy of curtailing bank credits and calling in loans in response to Jackson's reduction of federal deposits. The recession peaked in May. See Charles M. Snyder, *The Jacksonian Heritage: Pennsylvania Politics, 1833–1848* (Harrisburg, 1958), pp. 35–36. See list of names at the Manayunk meeting in *Pennsylvanian*, May 29, 1834.
3. *Pennsylvanian*, May 15, May 27, 1834.
4. Ibid., Oct. 6, Aug. 22, 1833; *Germantown Telegraph*, Nov. 27, 1833.
5. *Pennsylvanian*, June 3, 1834.
6. Ibid.
7. Tocqueville, *Democracy in America*, p. 220.
8. Seven farmers, five flour millers, two gentlemen, three innkeepers, and three teachers were in attendance (*Germantown Telegraph*, Sept. 3, Oct. 1, 1834).
9. See Laurie, *Working People*, p. 93; Snyder, *Jacksonian Heritage*, p. 48.
10. Episcopalian Charles Hagner and French Catholic Francis Barat were early members (*Germantown Telegraph*, Sept. 3, 1834).
11. Quote is in ibid., Aug. 6, 1834. Ferral and Catholic teacher Michael Sheehan met with Gowan personally to question him about his support of the Bank (ibid., Oct. 8, Oct. 15, 1834; Laurie, *Working People*, p. 93).
12. RBC, Minutes, Aug. 27, Sept. 14, 1833.
13. Quoted in Jones, *Roxborough Baptist Church*, p. 58.
14. Hobsbawm, *Primitive Rebels*, p. 140.
15. RBC, Minutes, Aug. 27, Sept. 14, 1833.
16. Ibid., April 21, July 19, 1834; Records of the Fourth Reformed Church, Manayunk.
17. RBC, Minutes, Oct. 25, Oct. 6, 1834; RBC, Register.
18. It is noteworthy that none of the other Philadelphia area Baptist churches reported that they had undertaken benevolent or moral reform (*Minutes of the Philadelphia Baptist Association*, 1835, HSP).
19. RBC, Minutes, March 19, Aug. 13, Sept. 19, Oct. 17, 1835.
20. *Germantown Telegraph*, May 20, 1835.
21. My interpretation of the makeup and function of the Flat Rock Temperance Society differs from that of Bruce Laurie, who views it as a self-conceived and self-directed workingmen's association (*Working People*, p. 51).
22. *Pennsylvanian*, June 18, 1835.
23. See John Ferral to Seth Luther in *The Radical Reformer and Workingman's Advocate*, July 18, 1835; Sullivan, *Industrial Worker*; pp. 134–37;

Cole, *Socialist Thought*, p. 128; Sidney Webb and Beatrice Webb, *The History of Trade Unionism* (New York, 1920), pp. 163–64; Laurie, *Working People*, p. 91. The textile operatives of the rural southern New England towns that Jonathan Prude has studied did not support the ten-hour movement of the 1830s or 1840s (*Industrial Order*, p. 225).

24. Ferral also informed Luther that "female labourers were about forming societies" (Ferral to Luther, June 22, 1835, in *Radical Reformer and Workingman's Advocate*, July 18, 1835). There is no evidence to indicate how many mills were struck. (*Germantown Telegraph*, June 24, 1835).
25. Quoted in *Radical Reformer and Workingman's Advocate*, June 27, 1835.
26. See letter from "J. F." in the *Radical Reformer and Workingman's Advocate*, July 4, 1835.
27. See ibid., Aug. 1, 1835. On the eve of the general strike in June 1835, the union was composed of twenty-three societies and had a combined membership of 7,000. Seven months later, forty-eight trade societies, with two alone numbering 900 members, belonged to the confederation (*Pennsylvanian*, Feb. 9, 1836; Sullivan, *Industrial Worker*, pp. 106, 149; see also Laurie, *Working People*, pp. 91–92).
28. Snyder, *Jacksonian Heritage*, pp. 52–60. The Jacksonian *Pennsylvanian* turned against Wolf because of his failure to carry out his pledge of opposition to the Bank. See *Pennsylvanian*, June 10, 1835.
29. *Germantown Telegraph*, Aug. 26, 1835.
30. Ibid., Aug. 12, Aug. 26, Sept. 9, 1835.
31. Ibid., Sept. 30, Oct. 14, 1835; Hazard, *Register of Pennsylvania* 16 (July–Jan. 1836): 277; and see Snyder, *Jacksonian Heritage*, p. 67.
32. *Germantown Telegraph*, Oct. 14, 1835; *Class Struggle and the Industrial Revolution*; pp. 108–9. See Wilentz's discussion of the demise of the workingmen's movement in New York (*Chants Democratic*, p. 215).
33. *Germantown Telegraph*, Oct. 5, 1836; Snyder, *Jacksonian Heritage*, p. 74.
34. *Germantown Telegraph*, Sept. 7, Aug. 17, Aug. 31, 1836.
35. Ibid., Sept. 7, 1836.
36. Ibid., Sept. 14, 1836. They also filed their own charges of political demogoguery against the politically ambitious Roxborough Whig, clerk Robert M. Carlisle. He was an active member of the Friends of Ireland and a long-time activist in the anti-Jacksonian movement in Roxborough. The Manayunk Whigs accused Carlisle of conspiring to exclude them from the delegate meetings.
37. It is difficult to believe, given the resolutions that were passed, that workingmen, particularly those who had worked for Rush, had voluntarily called the meeting, as the published proceedings stated. One also has to suspect the circumstances surrounding the election of Catholic Thomas Mullin, a gatekeeper on the canal, to the chair. Was he serving as a figurehead, coerced by economic pressure by one of the mill owners to promote the Manayunk Whigs' claim that they were an organization of workingmen? At the meeting of August 17 James Darrach, the Presbyterian mill owner, had presided as chairman. It is unlikely that Mullin would have willingly supported the resolutions that were passed at the meeting, given the fact that he had helped pen the anticapitalist proclamations against Ripka three years before.
38. *Germantown Telegraph*, Sept. 14, 1836.
39. Ibid., Sept. 28, 1836.

40. Ibid., Sept. 14, 1836.
41. See Sullivan, *Industrial Worker*, pp. 229–30.
42. On conflict that arose between mill owner Slater and the old elite of Dudley and Oxford, see Prude, *Industrial Order*, chap. 4.
43. *Manayunk Courier*, Feb. 12, 1848.
44. Scranton, *Proprietary Capitalism*, p. 249; Prude, *Industrial Order*, p. xii, and see pp. 142, 143, 225.
45. See Kulik, "Pawtucket Village, p. 6, and Wilentz, *Chants Democratic*, p. 143.
46. Foster, *Class Struggle and the Industrial Revolution*, pp. 252–53.
47. The quote is from Wilentz, *Chants Democratic*, p. 254, and see pp. 242, 243.
48. *Germantown Telegraph*, 28 Aug. 1836.

Bibliographic Essay

This essay is intended to identify the manuscript collections and published materials that proved the most important for writing the history of early industrialization in Philadelphia. I have been highly selective, and those interested in a comprehensive bibliography may want to refer to the dissertation from which this work evolved (University of California, Los Angeles, 1982). Because I was interpreting a broad social process, entailing economic, political, and cultural change, I used a large variety of sources. I regret that I could not uncover mill records for Manayunk's early period, which could have shed precious light on the experience of female mill workers. In general, the extant sources, quantifiable and literary, were dismally silent on women in the mills and the community of Roxborough.

Manuscript and Archive Collections

The John Nicholson Papers at the Pennsylvania State Archives in Harrisburg are a vast and rich collection of the business correspondence and records of this postrevolutionary manufacturing entrepreneur. Particularly important for my study were the Individual Business Accounts, 1787–1800, and the General Correspondence, 1772–1819. The latter is available in microfilm edition at the Van Pelt Library, University of Pennsylvania, and the Eleutherian Mills Historical Library, Wilimington, Delaware. This collection of letters received by Nicholson is complemented by the John Nicholson Letterbooks, 1791–1798, at the Historical Society of Pennsylvania, Philadelphia.

Two record groups in the Archives of the City and County of Philadelphia yielded vital evidence on structural change in Roxborough and the link between poverty and the textile industry in Philadelphia. The County Tax Assessor Ledgers and Duplicates for Roxborough Township, 1774, 1791, 1800, 1809, 1819, and 1822 (Record Group 1) were the key source of occupational and wealth data for the prefactory period. The records of the Guardians of the Poor (Record Group 35) reveal much about textile production and the daily work world of the dependent poor in early nineteenth-century Philadelphia. The Minutes of the Committee on Manufactures, 1807–1887, were particularly valuable. On textile production and workhouses in postrevolutionary Philadelphia one should also consult the Historical Society of Pennsylvania (HSP) for the records of the Pennsylvania Society for the Encouragement of Manufacture and the Useful Arts, Minutes of the Manufacturing Committee.

My analysis of church life and revivalism in the early 1830s depended heavily

Bibliographic Essay

on the following church records: Duplicates of the Records of the Fourth Reformed Church, Manayunk, 1831–1914 (HSP); Records of the Roxborough Baptist Church, which include meeting minutes and membership register, 1789–1858 (HSP); and the Historical Records of the Mount Zion Methodist Episcopal Church, Manayunk, at the Roxborough United Methodist Church, Roxborough.

A number of collections of personal papers and correspondence at the Historical Society of Pennsylvania proved extremely useful. The single most important was the Borie Papers. This collection includes the letterbooks, (1828–1834) of J. J. Borie, one of Manayunk's early textile mill owners. The Tench Coxe Correspondence and General Papers contain valuable evidence on the outlook of the promoters of textile manufacture in the new republic.

I relied heavily on the detailed Schedule of Manufactures, Fourth Census of the United States, 1820, for my analysis of the structure and circumstances of the textile trade on the eve of widespread mechanization. Weavers and manufacturers used the questionnaires to express their dismay over the state of depression in their trade.

Published Pamphlets and Government Documents

Two volumes in the British Parliament Papers series (Shannon, 1968) consist of hundreds of pages of testimony on the textile trades that indicate who was emigrating from Britain to America after the Napoleonic Wars and why. See *Emigration*, vol. 2 (1826–27), and *Industrial Revolution: Trade*, vol. 2 (1833). Mathew Carey pleaded the cause of the Irish immigrants and the urban laboring poor in Philadelphia in a series of pamphlets in the 1820s and 1830s. Particularly useful were *An Appeal to the Wealth of the Land, Ladies as Well as Gentlemen* (1833) and *Reflections on the Subject of Emigration from Europe, With a View of Settlement in the U.S.* (1826), both published in Philadelphia.

Statistics on the textile manufacture in Philadelphia in the 1830s can be gleaned from *Documents Relative to the Manufactures in the U.S.* (Washington, D.C.: U.S. Department of the Treasury, 1833). The testimony that provided the evidence of individual textile immigrants and reconstructed the social relations of the mechanized textile mill are contained in the Pennsylvania General Assembly record, *Journal of the Senate, 1837–38*.

Newspapers and Journals

My analysis of the mill strikes and political and social conflict in the early 1830s relied heavily on the *Germantown Telegraph, Pennsylvanian*, and *Mechanics' Free Press*. While the *Manayunk Courier* (1848) and *Manayunk Star* (1859–61) only started publishing in the years following my period of study, they contain useful stories about Manayunk's early years. These newspapers are all at the Historical Society of Pennsylvania. The two indexed weeklies, Samuel Hazard's *Register of Pennsylvania*, vols. 1–16 (1828–36) and H. Niles's *Niles' Weekly Register*, vols. 1–54 (1811–37) proved important for their cov-

BIBLIOGRAPHIC ESSAY

erage of Philadelphia economic news and local events not found in the newspapers.

LOCAL AND MANUFACTURING HISTORIES

Reminiscences and histories by Roxborough's nineteenth-century residents illuminated the details of Manayunk and Roxborough's development. Two were indispensable. Charles V. Hagner, who manufactured drugs in his canal mill and was an active participant in community politics and institutional life, recalled some of his experiences in *Early History of the Falls of Schuylkill, Manayunk, Schuylkill & Lehigh Navigation Companies, Fairmount Waterworks, etc.* (Philadelphia, 1869). A vast amount of miscellaneous data and statistics on early nineteenth-century Roxborough and Manayunk was compiled by resident Horatio Gates Jones in his "Early History of Roxborough and Manayunk" (a typescript of a lecture delivered before the Episcopal church in Manayunk in 1855) and "Historic Notes of Olden Times in Roxborough and Manayunk" (mounted articles from the *Manayunk Star*, 1859). I was able to recreate the history of the merchant millers of the Wissahickon from Douglas MacFarlan's detailed "The Wissahickon Mills" (1947), a three-volume scrapbook. All of these materials are at the Historical Society of Pennsylvania.

Much can be learned about the technology and organization of the early textile mills and the philosophy of the early manufacturers in British guides. See particularly James Montgomery, *A Practical Detail of the Cotton Manufacture of the United States of America*. (Glasgow, 1840), and Andrew Ure, *The Philosophy of Manufactures; or An Exposition of the . . . Factory System of Great Britain* (London, 1835). George S. White's *Memoir of Samuel Slater, the Father of American Manufactures* (Philadelphia, 1836) also details early textile manufacturing in America.

ARTICLES AND BOOKS

A wealth of evidence exists in the primary materials on attitudes toward manufacturing in the new republic. For an excellent analysis of these attitudes see Drew R. McCoy, *The Elusive Republic: Political Economy in Jeffersonian America* (Chapel Hill, 1980). See also Gordon C. Bjork, "The Weaning of the American Economy: Independence, Market Changes, and Economic Development," *Journal of Economic History* 24 (Dec. 1964): 541–60.

The literature on the development of textile manufacturing in New England is vast. The classic study on this topic remains Caroline F. Ware's *The Early New England Cotton Manufacture: A Study in Industrial Beginnings* (Boston, 1931). On the economic history of New England textiles, see Paul F. McGouldrick, *New England Textiles in the Nineteenth Century: Profits and Investments* (Cambridge, Mass., 1968). Two recent social histories taught me the most about the experience of textile workers and the nature of early factory society in New England: Jonathan Prude, *The Coming of Industrial Order: Town and Factory Life in Rural Massachusetts, 1810–1860* (Cambridge, Mass., 1983), and Thomas Dublin, *Women at Work: The Tranformation of Work and Community in Lowell, Massachusetts, 1826–1860* (New York, 1979).

Bibliographic Essay

The study of the development of textile manufacture and the early textile capitalists in Philadelphia should begin with the exhaustive study by Philip Scranton, *Proprietary Capitalism: The Textile Manufacture at Philadelphia, 1800–1885* (Cambridge, 1983). The classic case study of rural textile manufacture in the Philadelphia region is Anthony F. C. Wallace's *Rockdale: The Growth of an American Village in the Early Industrial Revolution* (New York, 1978). Two broader studies offer valuable insights into the history of Philadelphia's textile industry: David J. Jeremy, *Transatlantic Industrial Revolution: The Diffusion of Textile Technologies between Britain and America, 1790–1830* (Cambridge, Mass., 1981), and Carville Earle and Ronald Hoffman, "The Foundation of the Modern Economy: Agriculture and the Costs of Labor in the United States and England, 1800–1860," *American Historical Review* 85 (Dec. 1980): 1055–94.

I depended heavily upon secondary works for my discussion and analysis of British emigrants and emigration. I learned the most from William Forbes Adams, *Ireland and Irish Immigration to the New World from 1815 to the Famine* (New Haven, 1931); Maldwyn A. Jones, "Ulster Emigration, 1783–1815," in *Essays in Scotch-Irish History*, edited by E. R. R. Green (London, 1969); and Arthur Redford, *Labor Migration in England, 1800–1850* (Manchester, 1926). There is no major study on the experience of British immigrants in postrevolutionary Philadelphia. Two very informative articles on this subject are Edward Carter II, " 'A Wild Irishman under Every Federalist's Bed': Naturalization in Philadelphia, 1789–1806," *Pennsylvania Magazine of History and Biography* 94 (July 1970): 331–46, and Erna Risch, "Immigrant Aid Societies before 1820," *Pennsylvania Magazine of History and Biography* 60 (Jan. 1936): 15–33.

An unusually rich literature exists on the British laboring classes and oppositional thought in the early nineteenth century. The classic social history on this topic remains one of the most useful: J. L. Hammond and Barbara Hammond, *The Skilled Labourer, 1760–1832* (London, 1920). My own interpretation on the emergence of a working-class consciousness in industrializing society has been heavily influenced by two major works: John Foster, *Class Struggle and the Industrial Revolution: Early Industrial Capitalism in Three English Towns* (London, 1974), and E. P. Thompson, *The Making of the English Working Class* (New York, 1963). On the English radical economists, see G. D. H. Cole, *Socialist Thought: The Forerunners, 1789–1850* (London, 1952). An indispensable study of English trade union ideology and the English cotton spinners is R. G. Kirby and A. E. Musson's *The Voice of the People: John Doherty, 1798–1854; Trade Unionist, Radical, and Factory Reformer* (Manchester, 1975).

Philadelphia's own early nineteenth-century laboring classes are the subject of a handful of fine works. See William Sullivan, *The Industrial Worker in Pennsylvania, 1800–1840* (Harrisburg, 1955), Bruce Laurie, *Working People of Philadelphia, 1800–1850* (Philadelphia, 1980), and David Montgomery, "The Shuttle and the Cross: Weavers and Artisans in the Kensington Riots of 1844," *Journal of Social History* 5 (Summer 1972): 411–46. On Philadelphia's early labor movement see Lewis H. Arky, "The Mechanics' Union of

Bibliographic Essay

Trade Associations and the Formation of the Philadelphia Workingmen's Movement," *Pennsylvania Magazine of History and Biography* 76 (April 1952): 142–76. Two broader works contain valuable evidence on trade unionism in Philadelphia: John R. Commons et al., eds., *A Documentary History of American Industrial Society*, vols. 5 and 6 (New York, 1911; rpt., 1958), and Edward Pessen, *Most Uncommon Jacksonians: The Radical Leaders of the Early Labor Movement* (Albany, 1967). On the development of laboring-class thought in postrevolutionary Philadelphia, see the recent work by Ronald Douglas Schultz, "Thoughts among the People: Popular Thought, Radical Politics, and the Making of Philadelphia's Working Class, 1765–1828" (Ph.D. diss., University of California, Los Angeles, 1985).

In recent years, an important body of works has emerged on the formation of early industrial capitalist society and the working class in the antebellum Northeast. Besides the books by Prude, Dublin, and Laurie mentioned above, the most pertinent studies from this rich literature are Sean Wilentz's *Chants Democratic: New York City and the Rise of the American Working Class, 1788–1850* (New York, 1984), and Paul G. Faler's *Mechanics and Manufacturers in the Early Industrial Revolution: Lynn, Massachusetts, 1780–1860* (Albany, 1981).

Index

Adams, William F., 49
Allen, Leonard, 124
Almshouse. *See* Guardians of the Poor
Almy and Brown, 35, 36, 47
American Revolution, 142, 145
Anticapitalist thought, 143, 144, 155
Anti-Jackson politics, 130, 133, 150, 209n.36
Anti-Ricardian economists, 135, 147, 172
Appeal to the Wealthy of the Land, 61
Apportionment Act, 167
Arkwright spinning frame, 31, 32, 33, 35, 183n.33; and William Pollard, 8, 9, 24
Armstrong, David, 186n.65
Arson, 31, 116–17
Artisan Republicanism, 3, 135, 145, 172
Ash, Michael, 112

Baird, Isaac, 102
Baltimore, glassblowers from, 11
Bank of the United States, 124, 126, 155–57, 160, 173, 208n.11
Banks, 141, 142, 166
Bank war, 157, 158, 160
Baptist Church. *See* Roxborough Baptist Church
Baptists and temperance movement, 160–62
Barat, Francis, 108–9, 151–52
Batchelor, William, 106
Batley, John, 187n.64
Baxter, John, 47–48
Belfast, 34, 49
Benbow, William, 164
Beverly Cotton Factory (Mass.), 33
Biddle, Nicholas, 156
Bigony, John, 86
Binns, John B., 88
Blackburn (England), 118, 121
Blacksmiths, 124, 130
Blockley, 96, 97, 103, 146, 154
Boarding of workers, 72
Boatmen, 9, 11, 14, 17, 177n.18, 179n.36, 180n.53

Bolton, Thomas, 96
Bolton (England), 121
Boott, Kirk, 190n.29
Borie, J. J., 57–58, 62, 74–75, 92, 110, 171, 189nn. 12, 13, 15, 191nn. 36, 37, 41, 203nn. 22, 23; and local politics, 102, 158, 159; and mule spinners, 64–65, 120–22; and strike of 1828, 120–22, 137; and strikes of 1833–34, 135, 138, 150–53, 156, 158
Boston, 27, 49, 92, 164
"Boston Circular," 164
Bowker, John, 97
Bowker, Susan, 97
Bowker, Thomas, 97
Bowler, John, 23, 180n.55
Britain, immigrants from. *See* Immigrants, from Britain
Britain, textile manufacturing in, 28, 30, 33, 34, 57, 172; textile trades in, 28, 30, 34, 171; textile workers in, 30, 34, 48, 49, 67, 68, 140; trade movement in, ix, 1, 120, 135, 137, 143, 145, 147, 148
British Parliament, 9, 57, 63
Brook, William, 102
Brothers, Thomas, 165–66
Buenos Aires, 58
Button makers, 9, 11, 15, 16, 18, 19
Byrnes, William, 35, 184n.38

Campbell, John, 8, 10, 19–20, 23, 177nn. 8, 14
Canal workers, 11, 12, 108
Canton, 57
Capitalist class, 1, 93, 134. *See also* Industrial capitalist class
Capitalist class relations: 1, 4, 5, 6, 25, 54, 75, 129, 134, 145, 154, 155, 172; and anti-Ricardian economists, 135; and revivals, 137
Capitalist manufacturers, 7, 24, 25. *See also* Textile manufacturers
Capitalist manufacturing, 16, 24, 25

Index

Capitalist society, critique of, 118
Captain Swing, 117
Carey, Henry, 66
Carey, Mathew, 35, 60, 61
Carlisle, Robert M., 168, 209n.36
Carlyle, Robert, 144
Carpenters, 17, 90, 99, 121, 124, 130, 156
Carters, 9, 84, 95, 156
Catholics and Catholic Church in Manayunk, 108–9, 126, 137, 151–52, 156, 204n.47. *See also* St. John the Baptist Catholic Church
Centralized production, 7, 16; and control of workers, 29–32
Cheshire, 118
Chester County, 97
Child labor, 61, 64, 67–71, 138
Children: immigrant, 2, 56, 60, 61, 95–98, 102–4, 171; in strikes, 140–41, 154, 156, 164, 205n.17
Children in factories: in Europe, 28, 33; in Manauynk, 62, 63, 67–71, 88, 90, 96, 190n.25, 200n.7; in New England, 34, 48, 49, 186n.59; in Philadelphia, 27, 40, 42, 59, 61–63
China, 57
Churches, 4, 100, 104–10, 128–32, 160–62, 173; and strikes, 152–53. *See also* Revivals; *names of individual churches*
Clark, Joseph, 110, 207n.48
Class conflict, 3, 4, 132, 135, 139, 145, 154–55, 169–71. *See also* Labor and capital conflict
Class formation, 1, 3, 4, 170
Clay, Henry, 58
Claypoole, David C., 39
Cloth, importation of, 27, 28, 52, 57; types of, 57, 58, 65, 66
Cloth, market for, 30, 31, 47, 52, 55–57, 64, 171; in foreign countries, 57, 58, 75; and Guardians of the Poor, 41, 43, 46, 47
Combination Laws, 121
Competitive capitalism, critique of, 138, 139, 146, 155
Coopers, 84–87, 99, 101, 156, 196nn.27, 28, 29, 30
Cordwainers, 95, 121
Cotton, price of, 47, 119, 135, 138, 144
Cotton industry in England, 1, 33, 172. *See also* Textile manufacturing, in Britain
Cotton mills. *See* Mills
Cotton spinning machines, 23, 29, 30, 35–37, 41. *See also* Spinning jennies; Mule spinning; Spinning throstles
Cotton textile industry, 119. *See also* Textile manufacturing
Coxe, Tench, 7, 27, 28, 31–33, 36–37
Craig, Robert, 96

Crook, William, 137
Crosby, Joseph, 52

Danforth, Charles, 64
Danforth frame, 64, 65
Darrach, Cordelia, 107, 108, 131
Darrach, James, 93, 107–8, 131, 161–63, 166, 168, 209n.37
Darrach, Thomas, 93, 149
D'Arrusmont, Frances Wright, 144, 206n.27
Davenport, James, 184n.42
Davis, George W., 157
Davis, Robert, 109
Dean, Joseph, 103
Declaration of Independence, 145
Dedham, Mass., 97
Delaware and Schuylkill Canal Co., 11, 12
Democratic Press, 1, 55, 116, 117, 126
Democratic Republicans, 156–57, 166–67, 169. *See also* Jacksonians and Jacksonian party
Democratic Whigs of Manayunk, 159, 163, 167–69
Depression: following Napoleonic Wars, 51; following Panic of 1837, 134, 170; following Revolution, 7, 28
Discipline of workers, 12, 45
Doherty, John, 143, 147, 148, 153–54
Doran, Joseph M., 160
Dublin, Thomas, 74
Dublin, 49
Dudley, Mass., 204n.37, 210n.42
Dutch Reformed Church, 107, 131
Dyson, Martha, 152, 153

Early industrial capitalism, 3, 75, 116, 129, 145; critique of, 141–42, 144. *See also* Industrial capitalism
Early industrialization, 1, 31, 102. *See also* Industrialization
Early industrial society, 128, 133, 160, 173. *See also* Industrial capitalist society
East Indies, 39, 57
Economic clientage, 101
Economic elite of Roxborough, 5, 80, 85, 87, 92–94; and political authority, 100
Eichbaum, William, 14, 15, 19, 21–23
Elections: of 1828, 124; of 1830, 125–26; of 1832, 125, 128, 158; of 1835, 166; of 1836, 167–68, 170
Elkington, John, 125
Elouis, Henry, 13, 14, 15
Embargo Act of 1807, 46, 87
Embezzlement of materials, 29, 45–46, 186n.69
Engels, Frederick, 4
England, William, 8, 20, 23, 24
England, 76–77; cotton industry in, 1, 4,

[218]

33, 172; machine breaking in, 31, 118; textile mills in, 70–71, 96, 97; workhouses in, 38, 45
English, William, 165
English immigrants. *See* Immigrants, from England
English oppositional thought, 143–48, 155
Episcopal Church, 109–10
Equal rights, 3
Ervin, Edward, 52
Evangelical religion, 129–32, 137. *See also* Revivals
Evans, Oliver, machinery of, 77–79
Everman, Jacob, 86
"The Evil Effects of Labour-Saving Machinery Remediable," 118

Factories, 2, 27–31, 33. *See also* Children in factories; Mills; Textile manufacturing; Women in factories
Factory Regulation Act, 67, 71, 145
Factory rules, 73–74
Factory society, 5, 100, 110, 127
Factory system, 1, 4, 54, 66, 87, 163, 170–73; investigation of, 62; state regulation of, 71
Falls of Schuylkill, manufacturing at, 5, 7–25, 93
Family employment, 61–63, 95–98
Farmers in Roxborough, 4, 5, 80, 82–84, 87, 99, 116; and party politics, 127, 157, 159, 166–70
Farm labor, 9, 11, 83, 99
Federal Procession of 1788, 29, 30
Fertner, Mathew, 21
Female labor. *See* Women in factories; Women, laboring poor
Female labor and promotion of manufacturing, 32–33, 61
Female trades union, 165
Ferral, John, 138, 149, 152–54, 158, 160, 163–65, 206n.25, 208n.23, 209n.24; as J. F., 141, 145
First Presbyterian Church of Manayunk, 107–9, 128, 130, 131
Flat Rock Canal, 54, 87, 114
Flat Rock Temperance Society, 108, 162–63, 208n.21
Flax. *See* Linen, spinning of
Fleming, Joseph, 69
Fletcher, William, 186n.64
Flood, Thomas, 19, 23
Flour millers in Roxborough, 4, 76–79, 82, 85, 87, 93–94, 99, 101, 114–15, 194n.4, 199n.59; and politics, 127, 157, 159, 163, 166–70; and conflict over roads, 110–15, 194n.5
Flour mills in Roxborough, 5, 76–79, 84, 90, 195n.11
Flour trade, 76, 77, 79, 80

Foster, Israel, 97
Foster, John, 4, 172
Framesmiths, 9–11, 14–17, 45
France, James, 137
Freedly, Edwin T., 117
Freemasons of Roxborough, 203n.29
Friends of American Manufacturers, 27, 31
Friends of Ireland, 126, 160, 209n.36

Gallagher, Michael, 152
Gartside, Benjamin, 97
Gartside, James, 96, 97
General Ludd, 117, 118
General Strike of 1835, 135, 164–65
German immigrants, 76, 78, 95, 96, 108, 204n.47
Germantown, 67, 87, 100, 104, 111–15
Germantown Telegraph, 62, 72, 115–16, 119, 135, 160, 162, 164, 168; and strikes of 1833–34, 136, 139, 141–45
Gilmore, William, 138, 141, 147–48, 150, 152–54; and party politics, 156–57
Glasgow, 67, 71, 140
Glass manufacturing, 8, 20, 178n.27, 180n.53; at Falls of Schuylkill, 9, 14, 16, 20–21, 180nn. 47, 53
Glass workers, 9, 11, 14, 15, 20–22, 45, 180n.46
Globe Mill, 51
Good Intent Factory, 62
Gorgas, Samuel, 93
Gowan, James, 160
Graham, James Clark, 186n.61
Graham, Patrick, 105
Grand General Union of Cotton Spinners, 147
Gray, John, 146
Gruber, Joseph, 106
Guardians of the Poor, 37–46, 60, 66, 75, 184n.38, 184n.50, 185n.53
Gutman, Herbert, 15

Haddington, Penn., 47
Hagner, Charles V., 63, 90, 91, 92, 102, 103, 109, 114; and politics, 126, 157, 166; on strikes, 127–28, 140, 155, 208n.52; and testimony to Peltz Committee, 69, 70, 74, 88
Haines, Reuben, 67
Hamilton, Alexander, 8, 32, 33, 183n.25, 183n.28
Hamilton Company (Lowell), 71, 97
Handicraft production, 2, 16, 25, 26
Handlooms, 1, 29, 50, 51, 63, 65, 187nn. 82, 83
Handloom weavers, 1, 2, 26, 33, 37, 50–52, 116–19, 138, 154, 171; in Britain, 48–49, 118; and Guardians of the Poor, 40–44, 46, 185nn. 53, 54; 186nn. 61, 62,

[219]

Index

Handloom weavers (contd.)
65; independence of, 29, 63, 65–66; militancy of, 56, 64–66, 116–19, 202nn. 5, 7; and PSEMUA, 29, 30, 35, 181n.9; wage rates of, 52, 65, 66
Handloom weaving, 3, 26, 29, 31; after War of 1812, 48, 50–53; after 1820, 65, 66, 117
Hargreaves, James, 31
Haverford, Penn., 147
Haydock, Sam, 51
Haywood, Joseph, 109
Heighton, William, 118, 120, 123–24, 203n.16
Hibernian Society, 35, 60
Hinshilwood, John, 52
Hobsbawm, Eric, 132
Hodgkins, Thomas, 143, 147
Hours of labor. See Mills, hours of labor in
House of Industry, 60
Hudson, James, 150, 151, 161
Hughes, Daniel, 152
Hughes, William, 106

Ideology: of immigrant textile laborer, 3, 135, 143–46, 172; of laboring class, 142–43
Immigrant children. See Children, immigrant
Immigrants: from Britain, 2, 10, 34, 35, 45, 48–49, 171–73; from Britain, in 1820s and 1830s, 59–60, 63, 95–97, 145, 148; from England, 2, 10, 26, 35, 45, 48–49, 60, 171–73; from England, in 1820s and 1830s, 59–60, 63, 95–98, 199n.59; from Germany, 95, 96; from Ireland, 2, 10, 33, 35, 45, 49, 60, 117, 126, 171, 184n.38, 190n.19; from Ireland, in mills, 63, 95–96, 108, 199n.56; in Manayunk, 95–98, 127; from Scotland, 10, 36, 49, 59
Immigrant textile workers, 33–36, 48–49, 170–72; in 1820s and 1830s, 59–60, 63, 96–98, 127, 137; ideology of, 3, 135, 144–46, 172; in the mills, 87, 95–99, 144–45, 148, 199n.56; oppositional thought of, 3, 6, 135, 137, 144, 146, 148, 154–55, 172–73
Immigrant women, 2, 56, 60, 96–97, 171
Immigration to Philadelphia, 2, 10, 26, 34, 35, 48–49, 171; in 1820s and 1830s, 56, 59–60, 96–98
Incendiarism. See Arson
India, 57
Industrial capitalism, 123; critique of, 126, 206n.33. See also Early industrial capitalism
Industrial capitalist class, 138, 148, 159
Industrial capitalists, 87, 92, 127–28, 134, 139, 154, 155, 167, 169, 171
Industrial capitalist society, 1, 2. See also Early Industrial Society
Industrialization, 3, 5, 25, 26, 56, 116, 171, 172–73; in England, 4; in New England, 33, 55, 171; in Philadelphia, 54, 56. See also Early industrialization
Industrial production, 91, 116, 170; critique of, 135, 138, 171–72
Industrial system, critiques of, 67–68, 116, 118, 119, 146, 152–53, 155, 159; from England, 135, 143–46
Industrial working class, formation of, 1, 54, 95, 99, 134, 144, 145
Irish Catholics, 108–9, 126, 154, 159–60, 204n.47
Irish immigrants. See Immigrants, from Ireland

Jackson, Andrew, 156, 160
Jacksonians and Jacksonian party, 204n.35; in Philadelphia, 123, 150, 166, 167; in Roxborough, 6, 124–27, 134, 155–60, 166–67, 170, 173. See also Democratic Republicans
Jacksonville, 126, 127
Jaggers, George, 106
Jeffersonian Workingman, 72–73, 141–46, 173
Jenks, Alfred, 66, 117
Jeremy, David J., 187n.84
J. F. See Ferral, John
Johnson, Paul, 130
Jones, Amos, 127
Jones, Horatio Gates, 101, 102, 105, 110, 115
Joubert, Thomas, 180nn. 50, 53
Journeymen Cotton Spinners Society, 121

Kearns, Tony, 141
Keating, Eulalia, 108–9
Keating, Jerome, 58, 92, 93, 108–9, 120, 137, 138, 151, 204n.47
Kelly, Charles, 69
Kelly, George, 127
Kempton, James, 57, 63, 66, 117, 166, 168, 191n.39, 199n.59
Kennedy, Edward, 121
Kennsington, 54, 58, 98; weavers, 117
Koch, John, 109
Kuhn, Jacob, 137, 161–62

Labor and capital conflict, 2, 3, 64–66, 75, 116, 119, 127–28, 146, 164, 172; and court, 121–22; at Falls of Schuylkill, 13–15, 17, 19–22, 24–25; and party politics, 157, 166–67, 169, 173; and strikes of 1833–34, 138–41, 153, 156. See also Labor protest; Strikes

Index

Labor and capital, relations of, 66, 75, 127–28, 146, 153
Labor conditions in mills. *See* Mills, conditions of labor in
Labor, division of, 16, 196n.27
Laborers in Roxborough, 5, 84, 91–92, 95, 99, 116, 124; and politics, 125–26, 156–57; and revival, 130; and strikes of 1833–34, 138
Labor movement in Philadelphia, 4, 75, 120, 121, 136, 138, 146, 154, 172
Labor power, 16, 145, 146
Labor process, 3, 5, 15, 16; in mills, 67, 69, 70; of weaving, 29
Labor protest, 118; at Falls of Schuylkill, 16–17, 19–22, 25. *See also* Labor and capital conflict; Strikes
Labor radicals and radicalism in Philadelphia, 1, 67, 137–38, 145, 154
Labor-saving machines, 9, 30, 47–48, 55–56; and female and child labor, 32, 187n.74, 187n.81; opposition to, 117–19, 142
Labor, scarcity of, 10, 14, 32; in New England, 74, 193n.76
Labor supply, 32, 53, 61, 171
Labor theory of value, 142–43, 146
Laguerinne, Peter, 57–58, 92, 189n.13
Lancashire, 60, 71, 120; immigrants from, 95–97, 99, 109, 172; labor movement in, 137, 139, 142, 147, 149, 166, 205n.16; machine breaking in, 31, 118
Lancaster-Philadelphia Turnpike Co., 11, 12
Land prices in Roxborough, 82, 83
Laurie, Bruce, 137, 199n.56, 208n.21
Lebergott, Stanley, 75
A Lecture on Human Happiness (1825), 146
Lemon, James, 82
Levering, Abraham, 101, 200n.11
Levering, Anthony, 93–94, 198n.49
Levering, Charles, 127, 130, 161
Levering, John, 111, 200n.11, 201n.35
Levering, Michael, 101
Levering, Nathan, 80, 82, 101, 105, 113, 114, 200n.11
Levering, Peregrine (Perry), 94, 130
Levering, Samuel, 106
Levering, Wigard, 80
Levering, William, 80
Leverings, 80, 82, 87, 93–94, 101; and Baptist Church, 105–6, 110, 130, 205n.48; and party politics, 124, 127; and road building, 111–14
Leverington, 101
Lewis, David, 52
Linen, spinning of, 27, 28, 30, 36
Linn, George, 105
Lithgow, John, 8, 9, 10, 20, 23, 24

Liverpool, 49, 60
Livezey, Jonathan, 77–79, 113
Livezey, Joseph, 77–79, 94, 113
Livezey, Thomas, 79, 94
Livezeys, 78–80, 94, 101, 104, 113, 199n.4
London, 147
Londonderry, 34, 49
Loughrey, Margaret, 95–96
Loughrey, Neal, 95–96
Lowell, John Amory, 74
Lowell, Mass., 54, 55, 57, 99; labor and capital conflict in, 2, 135, 206n.28
Lowell mills, 58, 63, 66, 67, 74, 75, 96; and labor supply, 74–75; mill owners in, 92, 171; mill workers in, 175n.5, 193n.76; wages in, 71, 72, 192n.62
Luddites and Luddism, 117, 118, 172
Luther, Seth, 164, 208n.23
Lynn, Mass., 3, 135, 206n.28

McCurdy, Charles, 185nn. 54, 57
McGlinchy, Patrick, 137
Machine breaking, 3, 31, 32, 116–19
Machine makers, 8–11, 23, 29, 45, 130
Machinery. *See* Textile machinery
Machinists, 90, 95, 124
McIntyre, William, 151
Managers, 25, 74, 92
Manayunk, 55, 119; burning of mill in, 116–17; Catholics and Catholic church in, 108–9, 126, 137, 151–52, 156, 204n.47; and comparison to Manchester, 54, 55, 66, 67, 127, 144, 170, 171; emergence of factory system in, 1, 54, 55, 78, 99, 116, 170; families employed in, 62, 63, 73, 89, 90, 95–98, 104; growth of, 88, 90–91, 93, 114, 197n.40; immigrants in, 95–98, 127; map of, 89; occupational groups in, 90, 91; political conflict in, 4, 116, 156–60; 163, 167–70, 173; population of, 88, 90, 91, 95; social conflict in, 75, 116, 127–28, 172; town council, 100, 102. *See also* Mill owners in Manayunk; Mills, size and number in Manayunk; Mill workers in Manayunk
Manayunk Rifle Company, 126
Manchester, England, 118; comparison to Manayunk, 54, 55, 66, 67, 127, 144, 170, 171; immigrants from, 96, 97; labor conflict in, 118, 120–21, 134, 137, 139, 147; labor radicalism in, 143–44, 147–48, 207n.41; mills of, 67, 70, 71, 170
Manchester New Lights, 144, 145
Manufactories at Falls of Schuylkill, 7–9, 12, 16, 18, 20, 25
Manufactory production: in Philadelphia, 2, 5, 7; system of, 16, 24, 26, 66
"Manufacturing aristocracy," 134, 159, 169, 173

[221]

Index

Manufacturing Census: of 1810, 36–37; of 1820, 50–52
Manufacturing, promotion of, 7, 8, 27, 28, 33, 187n.82
Markel, Jacob, 112
Market, 55–59, 75. *See also* Cloth, market for
Marshall, John (spinner), 121, 123
Marx, Karl, 3, 16
Masons, 90, 124, 130
Massachusetts: mills in, 35, 74–75; textile workers in, 99
Matsen, Ann, 152, 161
Maxwell, James, 52
Mechanic-manufacturers, 8, 23–25
Mechanics' Free Press, 55, 62, 67, 98–99, 118, 123, 126, 146
Mechanics ideology, 142
Mechanics' Union of Trade Associations (MUTA), 118, 120, 123, 148
Mechanization: of mills, 66, 67; in New England, 2, 33–35, 48, 50, 171; in Philadelphia, 2, 3, 33, 34, 48, 53–55, 56, 60, 171; of spinning, 29, 32, 47–48, 63–65, 171; of weaving, 48, 63, 65, 66
Mendenhall and Cope, 30
Merchant capitalists, 25
Merchant millers. *See* Flour millers in Roxborough
Merwine, Andrew, 86
Methodism, 107
Methodist Church, 106–7
Mexico, 57
Miles, Benjamin, 124, 130
Militancy of textile workers, 56, 64–66, 99, 165, 172–73. *See also* Strikes
Militia system, 141, 142, 165, 166
Miller, Lawrence, 86, 101, 196n.27
Miller, Roberta Balstad, 196n.23
Mill owners, 57, 171; and attempts to mechanize, 63–66; in New England, 57, 74–75, 171
Mill owners of Manayunk, 4, 55, 56, 58, 75, 99, 110; and churches, 107–10; cooperation among, 122, 149; as economic elite, 92, 93, 114–15; and labor supply, 62, 63, 95; and market, 57, 58, 119; and politics, 100, 102, 114–15, 156–57, 159, 163, 166, 167–69, 171; and public schools, 102–3; and strikes of 1833–34, 141, 144–45, 151, 154–55, 173; and wage reductions, 120–22, 132, 135, 139, 144; and work rules, 73–74
Mills, 1, 5, 54, 55, 57, 67, 87, 97, 165, 170, 171; burning of, 116–17; conditions of labor in, 66, 67–71; critique of, 138–39, 142, 154, 171; hours of labor in, 67–69, 71, 154, 163, 164. *See* Lowell mills, males employed in, 91, 92; in Massachusetts, 35, 74–75, 99; in New England, 67, 97; in Pennsylvania, 63; in Rhode Island, 35, 36–37, 39; rules in, 73–74, 193n.70; size and number in Manayunk, 54, 55, 58, 88, 90; social relations of, 54, 127, 129, 170, 173; wage rates in, 71–73, 192n.62, 193n.64
Mill strikes. *See* Strikes in mills
Mill towns in New England, 2, 97, 172, 208n.43
Mill workers: diseases and disorders of, 70; immigrant, 95–99, 170, 172, 205nn. 3, 7; and general strike, 164; and labor movement, 135, 138, 172–73; mobility of, 96–99, 132, 193n.69, 199n.59; in New England, 63, 74–75, 172, 188n.87, 190n.29; and oppositional thought of, 142, 175n.7; in Philadelphia, 4, 119, 169, 172, 207n.43; and politics, 124, 156–59, 173; and revivals, 130–33, 161; in Rhode Island, 62, 74, 193n.76; and strikes of 1833–34, 136, 138–40, 146–48, 150–56, 173
Mill workers in Manayunk, 4, 9, 71, 75, 87–91, 116, 123, 154, 172; immigrant, 95–98; numbers of, 55, 61–63; 172–73. *See also* Mill workers
Millwrights, 84, 90
Mix, Jonathan, 18, 19, 23, 179nn. 39, 43
Mobility, 84, 86, 87; of mill workers, 96–99, 132, 193n.69, 199n.59
Montgomery, David, 204
Montgomery, James, 71
Moose, George, 87
Moral economy, 139
Morris, Robert, 8
Mosely, Thomas, 96, 97, 192n.60
Mount Zion Methodist Episcopal Church, 106–7
Moyamensing, 51, 52, 54
Moyer, George, 127
Moyer, John, 127
Muhlenberg, Henry A., 165, 166, 167
Mules in Manayunk mills, 55, 64, 65
Mule spinners, 156, 171, 191n.33; employed by Guardians of the Poor, 40–42, 45, 185n.53, 186n.64; in Manayunk mills, 63–65, 70, 142; militancy of, 56, 63–65, 121–23, 137, 147; mobility of, 97; numbers of in Manayunk, 90, 95; in Philadelphia, 1, 2, 26, 33, 36, 50, 51, 61, 63–65, 171; in strike of 1828, 120–22, 123, 137; in strikes of 1833–34, 135–37, 140–41, 152, 154; wages of, 64
Mule spinning, 36–37, 50, 63–64, 191n.33
Mullin, Thomas, 137, 209n.37
Murphy, Francis, 109, 124

Napoleonic Wars, 59, 84, 147
Nash, Gary B., 27
Nashua Company, 64

[222]

Index

National Association for the Protection of Labor, 147
National Republicans, 6, 125, 127, 161. *See also* Whigs and Whig Party
National Trades Union, 154
Natural rights, 138, 142, 146
Neld, Luke, 96, 199n.59
Nelson, Andrew, 44
New England: employment of women in, 49, 50, 74–75; industrialization in, 33, 35, 171; mechanization in, 2, 33–35, 48, 50, 171; mill owners in, 57, 74–75, 171; mills in, 67, 97; mill towns in, 2, 97, 172; mill workers in, 63, 71, 172, 174–75, 190n.29; textile manufacturing in, 2, 5, 26, 29, 33, 48, 50, 72; trade union movement in, 148–49
New Hampshire, 64
New Jersey, 62, 63
Newry, 34, 49
New York (City), 175n.1, 179n.37, 181n.2, 203n.26, 204n.32; immigration to, 35, 49, 60; labor and capital conflict in, 3, 135, 206n.33; trade union movement in, 206n.33, 207n.35
New York (state), 10, 77, 78
Nichols, Reverend Dyer A., 110, 129, 130, 161, 162
Nicholson, John: and manufacturing at Falls of Schuylkill, ix, 5, 7–25, 45, 54, 66, 75, 177nn. 5, 8, 13, 14, 179nn. 36, 39, 180nn. 47, 53, 55, 56, 58; and PSEMUA, 26, 28
Niles' Weekly Register, 49, 57, 117, 187n.81
Norristown, 120, 132, 147, 169, 205n.17
The Northern Liberties, 54, 117, 124

Observer, 141, 143–45, 206nn. 25, 26, 27
Occupational groups in Roxborough and Manayunk, 83, 84, 90, 91
Ogden, Sally, 153
Ogden, Samuel, 72, 74, 96, 97, 98, 150, 153, 158
Outwork production in Philadelphia, 2, 26–28, 32, 53, 54, 66, 171; of Guardians of the Poor, 40–41, 43–46
Owen, Robert, 144, 206n.24
Owenite cooperativism, 148, 206n.24
Oxford, Mass., 204n.37, 210n.42

Paine, Thomas, 144
Panic of 1837, 134, 170
Papermakers, 101
Papermills, 78, 84, 90
Party politics. *See* Political parties
Passenger Act of 1803, 35, 184n.38
Paternalism, 100, 101
Paterson, N.J., 8, 10, 64
Patterson, Elizabeth, 131, 132

Patterson, Mary, 131, 132
Paul, Joseph, 39
Pawtucket, R.I., 2, 55, 66, 179nn. 34, 38, 190n.29
Peltz Committee, 62, 66, 68–71, 73, 103
Pennsylvanian, 126, 136, 140, 146, 152, 156, 157, 158, 209n.28
Pennsylvania Senate, Factory Investigation Committee. *See* Peltz Committee
Pennsylvania Society for the Encouragement of Manufacture and the Useful Arts (PSEMUA), 31, 32, 38, 181nn. 6, 9; and manufacturing of textiles, 7, 27–30
Perceiver, 141, 144, 145
Philadelphia: as entrepôt, 47, 57, 78; as immigrant port, 2, 26, 34, 35, 48–49, 59–60; labor movement in, 4, 75, 120–21, 136, 138, 146, 154; mechanization in, 2–4, 53, 54, 56, 60, 171; unemployment in, 7, 32, 59, 66, 119; workhouses in, 27–29, 32, 171. *See also* Spinners in Philadelphia; Textile manufacturing in Philadelphia; Weavers in Philadelphia
Philadelphia Baptist Association, 128
Philadelphia, Germantown, and Norristown Railroad, 91, 115
Philadelphia *U.S. Gazette*, 151, 164
Phillips, Amos, 157
Piecers, 61, 70, 96
Piece work, 17, 27, 43, 179n.36. *See also* Outwork production in Philadelphia
Pike Creek, 147
Pittsburgh, 10, 23, 62
Political economy and economists, 143, 144
Political parties, 123–28, 156–60, 163, 165–70, 173
Politics in Roxborough. *See* Roxborough, politics in
Pollard, Sidney, 38
Pollard, William, 8, 10, 12, 14, 18, 23, 24, 176n.3, 177nn. 5, 13, 179nn. 37, 38, 180nn. 55, 56
Poor, employment of: in factories, 33, 45, 59–63, 171; by Guardians of the Poor, 39–41, 44, 45; by PSEMUA, 27–29, 32
Poor in Roxborough, 92, 101
Population: of Manayunk, 88, 90, 91, 95; of Roxborough, 80, 84, 88, 90, 91, 95
Porter, John, 106
Poverty, 7; of immigrant workers, 2, 5, 35, 53, 60, 61, 99, 127; in Roxborough and Manayunk, 91, 92, 99; and textile production, 26, 28, 59–62
Powerloom, 50, 191n.39
Powerlooms in Manayunk, 55, 56, 58, 63, 65, 66; attack of, 116–17
Powerloom weavers, 1, 62, 65, 66, 122, 150, 171; mobility of, 97; and strikes of 1833–34, 140–41; wages of, 65, 71, 72

[223]

Index

Powerloom weaving: effects on handloom weavers, 50, 65–66, 117–19; in Philadelphia, 50–52, 63, 65, 66
Presbyterian church. *See* First Presbyterian Church of Manayunk
Presbyterians: and revivals, 128, 130–31, 161, 162, 204n.47; and temperance movement, 160, 162, 163
Primitive Methodists, 161
Principles of Political Economy (1817), 135
Producing class, 123–24, 135, 139, 142–43
Proletarianization, 3
Property holding in Roxborough, 81–87, 91, 94
Propertylessness in Roxborough, 80–81, 84–86, 91, 99, 197nn. 42, 43
Protestants, 126, 137, 154, 157, 160, 204n.34, 204n.47
Providence, R.I., 49, 57
Prude, Jonathan, 172, 179n.34, 181n.1, 190n.29, 199n.61, 204n.37, 208n.23
PSEMUA. *See* Pennsylvania Society for Encouragement of Manufacture and the Useful Arts
Public education, 125, 165, 166
Public schools, 100–104
Puttingout System. *See* Outwork production in Philadelphia

Quaker settlers, 76, 78, 104
Quarrymen, 9, 11, 12, 14, 15, 17

Radical economic theory and theorists, 3, 135, 143, 146, 147–48, 172, 205n.3
The Radical Reformer and Workingman's Advocate, 165
Railroads, 91
Rawley, William, 130, 161
Recessions, 119, 149
Redford, Arthur, 45
Registry Law, 167
Relations of Production, 25, 66, 84, 87
Religion, 104, 105–10, 116, 128–33
Religious revivals. *See* Revivals
Renshaw, James, 124, 148–49
Republican sentiments, 125, 139, 145
Revivals, 4, 5, 6, 110; and class relations, 129, 137; of 1832–33, 128–33, 137, 150, 153, 160–62, 170, 173
Rhode Island: mills in, 35, 36–37, 39; mill workers in, 62, 74, 193n.76
Ricardo, David, 135, 143
Richards, Mark, 122, 149, 203n.22
Ridge Road, 77, 79, 93, 111–12, 114–15
Ridge Turnpike, 83, 87, 114
Righter, Michael, 111, 201n.35
Ripka, Joseph, 58–59, 62, 68, 92, 93, 109, 119, 142, 171, 189n.14; and discharging of employees, 74, 158; employees of, 96–98, 103, 150, 192n.60, 199n.59; and mills attacked, 117; and politics, 102, 158–59, 166, 168; and strikes of 1833–34, 135, 138, 141, 146, 148, 209n.37
Ritner, Joseph, 166
Rittenhouse, Abraham, 79, 112
Rittenhouse, Enoch, 94, 114
Rittenhouse, Henry, 78, 111
Rittenhouse, Jacob, 112, 159
Rittenhouse, Martin, 79, 94, 111–13
Rittenhouse, Nicholas, 78, 101
Rittenhouse, Nicholas, Jr., 94, 127, 159, 167
Rittenhouse, Wilhelm, 78
Rittenhouse, William, 79
Rittenhouses, 78–80, 94, 101, 104, 106, 199n.4; and party politics, 127, 159, 167; and road building, 111–15
Rittenhousetown, 84, 101
Roach, Michael, 45
Road building, 110–15, 201n.35
Robeson, Andrew, 79
Robeson, Jonathan, 77, 78, 159
Robeson, Peter, 79, 83, 86, 113, 194n.9, 195n.19, 199n.4
Robesons, 78–80, 101
Rochefoucauld-Liancourt, Francis A. de La, 9, 13, 77, 78, 82, 83
Rochester, N.Y., 107, 129, 130
Rockdale, Penn., 73, 97, 98, 99, 129, 192nn. 46, 62, 200n.7
Rosenberg, (estate), 92, 109
Roseville, Penn., 147
Roxborough, 7; distribution of wealth in, 80–82, 84–86, 92; economic elite of, 5, 80, 85, 87, 92–94, 100; economy of, 5, 87, 91, 92, 95, 100; as flour producing area, 76–78, 87; and growth of Manayunk, 54, 55, 87, 88, 95, 102; map of, 89; occupational groups in, 83, 84, 90, 91; political conflict in, 4, 95, 115, 156–60, 163, 167–70, 173; political and cultural institutions in, 100, 102, 104, 109; politics in, 100–102, 111, 116, 123–28, 156–60, 167–70, 173; population of, 80, 84, 88, 95; and religious awakening in, 128–32; social conflict in, 4, 100, 116, 134, 159, 170, 173; social groups in, 5, 76, 82, 83, 87, 95, 100, 173. *See also* Churches; Manayunk; Political parties; Property holding in Roxborough; Propertylessness in Roxborough; Strikes in 1833–34
Roxborough Baptist Church, 105–6, 110, 124, 200n.11; and revival, 128–32, 137, 160–62, 205n.48; and strike, 161–62; and temperance movement, 160–62
Roxborough Workingmen's Party, 123–27, 166
Rudolph, Sebastian, 96

Index

Rush, John, 159, 166, 168, 209n.37
Ryan, Mary, 205n.50

Sabbath schools, 100, 104, 200n.9
St. Andrews, 49
St. David's Episcopal Church, 109–10
St. John the Baptist Catholic Church, 95, 108–9, 137, 198n.54
Sanders, William, 196n.30
Schlesinger, Arthur, M., Jr., 126
Schofield, Mary Ann, 74
Schools. *See* Public schools; Sabbath schools
Schultz, Ronald, 205n.3
Schuylkill Factory, 92, 93, 108; strike in, 120–22, 138, 150–53
Schuylkill Navigation Co. and canal, 1, 54, 55, 72, 84, 114
Schuylkill River, 54, 76–77, 80, 82, 87, 93, 111, 113, 114
Scotland, 60
Scott, Joan, 178n.27, 180n.46
Scranton, Philip, 172, 187n.77, 189n.44, 197n.43, 198n.46, 202n.2, 205n.7
Shaw, William, 199n.59, 200n.9
Shoemakers, 138, 152, 161
Shopkeepers, 4, 5, 108, 110, 119, 124; and politics, 156–57, 170; and revival, 128, 130, 170; and strikes of 1833–34, 135, 137, 154, 155
Silesia Manufactory, 58, 97
Simpson, William, 161
Singer, William, 161, 162
Skilled labor, ix, 10, 16, 25, 29
Slater, Samuel, 35, 55, 66, 179n.34, 191n.39, 210n.42
Slater Company, 57
Small, William F., 138, 147, 150
Smick, George, 93
Smith, Samuel, 110
Snyder, Conrad, 44
Snyder, Michael, 157
Social conflict, 4, 53, 100, 116, 118, 134, 159, 170
Social reform, 130
Society of Mechanics and Workingmen, 165
Southern New England, 2, 63, 71, 92, 171–72, 208n.23
Southwark, 138, 160, 166
Spinners, 2, 3, 5; in Manayunk, 4, 64–65; in Philadelphia, 4, 15, 27–29, 50–51, 53, 63, 64, 171. *See also* Mule spinners; Throstle tenders
Spinners, female, 33, 37; of Guardians of the Poor, 39–41, 44–46, 185n.53; and PSEMUA, 27–32
Spinning: mechanization of, 29, 32, 47, 48, 53, 63–65, 171; in workhouses, 27–31

Spinning jennies, 29–31, 36, 51, 182n.15
Spinning throstles, 50, 55, 56, 63–65, 191n.33
Spinning wheels, 1, 27–30, 37, 39, 41
Steam engines, 8
Stocking makers, 9, 15, 19, 20
Stocking manufacturing, 8–10, 16, 24
Stockport, England, 121
Store, John, 44
Stranghan, James, 52
Strike breakers, 121, 151, 152, 205n.17
Strike committee of the Schuylkill Factory, 150–53, 158, 207n.41
Strikes: in mills, 2, 3, 4, 6, 65, 75, 109, 120, 133, 169, 172; in 1828, 120–22; in 1833–34, 134–44, 146–56, 161, 171, 207n.42. *See also* General strike; Labor and capital conflict; Labor protest
Stumme, George, 7, 177n.5
Sullivan, William, 4, 122
Supervisors: conflict with John Nicholson, 18, 19, 22–23, 24; of Guardians of the Poor, 41, 42, 46; and Nicholson's workers, 8, 12, 13, 18, 19, 23
Surplus value, 143
Sutherland, Joel B., 160
Sweetman, John, 21

Tariff: of 1816, 51–52; of 1832, 136; as political issue, 141, 142
Taylor, Charles, 8, 19, 23–24, 180n.58
Tax assessments in Roxborough, 80, 81, 82, 84, 86, 90
Technological improvements in machinery, 56, 58, 63, 64, 75
Temperance movement, 134, 160–63, 170
Tenants and tenancy, 82, 83, 101
Ten-hour movement, 163–65, 208n.23
Ten-hour system, 149, 163, 164
Textile factories, promotion of, 32, 33
Textile machinery, 10, 29, 36–37, 152–53, 171, 188n.86; in Manayunk, 54, 55, 64, 65, 66
Textile manufacturers, in Manayunk. *See* Mill owners, in Philadelphia, 4, 5, 32, 51–52, 61, 171, 188n.6
Textile manufacturing: in Britain, 1, 28, 30, 33, 34, 52, 172; by Guardians of the Poor, 37–46, 60, 66, 75, 184n.38, 184n.50, 185n.53, 186n.65; in New England, 2, 5, 26, 29, 33, 48, 50, 172; in Philadelphia, 2, 5, 26, 27, 47, 50–54, 171–72; and the poor, 33, 37, 60–63; in Rhode Island, 62; in workhouses, 26–29, 32, 38
Textile operatives. *See* Mill workers
Textile workers: in Britain, 30, 34, 48, 49, 67, 68, 140; female, 27–30, 61–64, 70, 71; in Massachusetts, 94; in Philadel-

[225]

INDEX

Textile workers (contd.)
  phia, 2, 5, 47, 60, 62–63, 188n.6. See also Mill workers; Women in factories
Thompson, E. P., 4, 15, 63, 107, 116, 132
Thompson, William, 143
Thornily, John, 63, 69, 98, 192n.51
Throstle tenders, 1, 64, 71
Tocqueville, Alexis de, 134, 159, 169
Tradesmen, 4, 5, 84, 91, 92, 101, 109–10, 116, 119; and politics, 123–25, 127, 156–57, 169–70; and revival, 128, 130, 170; and strikes of 1833–34, 135–36, 140, 143, 150, 154–55
Trade unionism, 173; in Britain, 1, 3, 135, 137, 143, 145, 147–48, 155, 172
Trade unionists, 135, 155, 206n.33
Trade union movement: in Britain, 120, 135, 137, 143, 145, 147–48; of factory operatives, 5, 120, 135, 138, 146–49, 154, 163, 172–73; in New England, 148–49; in New York, 206n.33, 207 n.35; in Philadelphia, 136, 138–39, 147, 154, 170
Trades Union of the City and County of Philadelphia (TUCCP), 148, 165, 209n.27
Trades Union of Pennsylvania (TUP), 146–47, 149–50, 158
Treillou, J. J., 121, 123
Tufnell, Henry, 155

Ulster, 49
Unemployment in Philadelphia, 7, 32, 59, 66, 119
Union Benevolent Society, 119
United Company of Philadelphia for Promoting American Manufacture, 27, 37–38
United Society for Manufacturers, 27
Unskilled labor, 9, 11, 59, 91, 95. See also Laborers in Roxborough
Usher, A. P., 73
Utica, N.Y., 129, 205n.50

Virginia, 77–78

Wage-labor system, 1, 2, 5, 171; in manufactory, 15, 16, 17, 21, 24–25
Wage payments: conflict over, 17–22, 25; forms of, 18–19, 22
Wage rates: of Guardians of the Poor workers, 43, 185n.53; of handloom weavers, 52, 65, 66; in mills, 71–73, 142, 153, 179n.34, 192n.62, 193n.63; of mule spinners, 64, 72, 73; of Nicholson's workers, 10, 11, 15, 17, 22; of Roxborough trades, 87; unskilled labor, 9, 11
Wage reductions, 119–20, 122, 128, 147–49, 169, 171, 172; as political issue, 124, 127, 173; and strikes, 135–36, 138–39, 141–44, 150–51, 153–54
Wages, as property rights, 146, 150–51
Wagner, Samuel, 93, 109
Wagner, Tobias, 93, 109
Wakefield, 140
Walker, Samuel, 186n.65
Wallace, Anthony, 97, 98, 192nn. 46, 62, 200n.7, 203n.14, 207n.41
Waltham system, 2
War of 1812, 46, 48, 84, 87
Water power, 76; sale of, 54, 55, 87, 93
Wealth distribution in Roxborough, 80–82, 84–86, 92
Weavers, 2, 5, 171; in Manayunk, 4, 95; in Philadelphia, 4, 47, 50–53, 60, 171. See also Handloom weavers; Powerloom weavers
Weaving. See Handloom weaving; Powerloom weaving
Welsh, William, 157
West Indies, 39, 57, 76, 79
Wetherill, Samuel, 28, 184n.42
Whigs and Whig Party: in Pennsylvania, 167; in Philadelphia, 160; in Roxborough, 134, 157–60, 166–70, 173. See also Political parties
Whitaker, Henry, 50, 61, 193n.67
Whitaker, William, 97
White, James, 96
Whitesmiths. See Framesmiths
Wilentz, Sean, 175n.1, 179n.33, 203n.26, 204n.35, 206n.29
Wilkinson, Jeremiah, 96
Willis, Beriah, 130
Wilmington, Del., 36, 179n.37
Winpenny, John, 116
Wise, John, 112–13
Wissahickon Creek, 7, 82, 84, 115; flour mills on, 76–80, 111, 113
Wolf, George, 165, 166, 167, 209n.28
Wolf, Stephanie, 87
Women: employed by Guardians of the Poor, 39–41, 43–46, 185nn. 53, 57; as heads of households, 92, 106; immigrant, 2, 56, 60, 61, 97, 171, 190n.19; laboring poor, 27, 28, 59–63; in Roxborough churches, 106–7, 131–32, 161, 162; in strikes, 140–41, 152–54, 156, 161, 164, 165; in workhouses, 27–30, 181n.9, 190n.23
Women in factories: in Europe, 28, 33; in Manayunk, 90, 91, 96–97, 203n.23; in New England, 49, 50, 74–75, 184n.45; in Philadelphia, 61–64, 70, 97, 171, 175n.5, 193n.64
Wool, spinning of, 27
Workers: customary habits of, 16, 22; drinking among, 14–16, 178n.28, housing of, 8–9, 13, 88; in manufactories, 5

[226]

Workhouses: in England, 38, 45; in Philadelphia, 26–29, 53, 75, 171
Working class: political awakening of, 128, 144, 146, 158–59, 161; and politics, 166–67, 169–70; and revival, 129
Working-class consciousness, 2, 3, 99, 135, 139, 145–47, 154, 155, 172, 173; in New York, 206n.33
A Workingman. *See* Jeffersonian Workingman
Workingmen of Manayunk, 124
Workingmen's Party: of New York, 203n.26; of Philadelphia, 123, 165; of Roxborough, 6, 123–27, 150, 166, 170, 203n.29

Working People of Manayunk, 68, 135–36, 138, 139, 142, 146–47, 154, 161
Wright, Fanny, 144. *See also* D'Arrusmont, Frances Wright

Yarn: coarse, 9, 26, 28, 31, 36, 64; fine, 2, 9, 26, 36, 53, 61, 64, 171; market for, 31, 36, 44, 47, 64
Yorkshire, 60; immigrants from, 95–97, 109
Young, William, 90
Young, William T., 124, 150, 152
Young Men's Missionary Society, 107

THE MILLS OF MANAYUNK

Designed by Ann Walston

Composed by Harper Graphics, Inc., in Trump

Printed by Thomson-Shore, Inc.,
on 50-lb. Warren's Olde Style Wove
and bound in Joanna Arrestox A